ATTRACTION, LOVE, SEX

ATTRACTION, LOVE, SEX

THE INSIDE STORY

SIMON LeVAY

Columbia University Press   *New York*

Columbia University Press
Publishers Since 1893
New York   Chichester, West Sussex
cup.columbia.edu

Library of Congress Cataloging-in-Publication Data
Names: LeVay, Simon, author.
Title: Attraction, love, sex : the inside story / Simon LeVay.
Description: New York : Columbia University Press, [2023] |
Includes index.
Identifiers: LCCN 2022035515 (print) | LCCN 2022035516 (ebook) |
ISBN 9780231204507 (hardback) | ISBN 9780231555579 (ebook)
Subjects: LCSH: Sex (Psychology) | Sex (Biology) | Sex.
Classification: LCC BF692 .L465 2023 (print) |
LCC BF692 (ebook) | DDC 155.3—dc23/eng/20221212
LC record available at https://lccn.loc.gov/2022035515
LC ebook record available at https://lccn.loc.gov/2022035516

Cover design: Catherine Casalino

CONTENTS

PREFACE

In 2021 alone, according to Google Scholar, over 200,000 newly published academic articles and books included the words "sex" or "sexual" in their titles. Keeping up with this flood of material, let alone with the countless articles and books that cover sexual matters without using those terms, would require reading one publication every two minutes, with no time off for eating, sleeping—or sex.

Lacking that degree of fortitude, I've decided to pick a few topics out of the general field of sexuality and a few studies within each field that illustrate the overall message I want to convey: that sex, in spite of its sexiness, is an appropriate subject for rational inquiry just like any other aspect of human nature; we can learn something interesting and important about ourselves by viewing sex through a scientific lens.

Things that may seem obvious, like why we have sex in the first place, become unobvious when we look at them closely. It has taken clear thinking, clever experimentation, and vigorous debate to find answers. Many questions are not yet fully resolved; some remain complete mysteries.

I'm lead author of a college textbook of human sexuality. What I've found in the course of working on that text and from lecturing is that both students and instructors want decisive answers—facts, in other words, that lend themselves to the setting or sitting of exams. This book, though, is not a textbook, and so I feel no need to shy away from the uncertainties and loose ends. Readers may sometimes find this incompleteness frustrating, but that's how science works. Scientific progress is like the slow retreat of a morning fog, when advancing sunlight opens up some vistas while leaving others in obscurity. Perhaps a few readers will be spurred to join the scientific enterprise: to answer one of the unanswered questions or to pose a question that no one has yet thought to ask.

To those readers who are already engaged in sex research, I apologize if I don't do full justice to your work, if I credit it to your lab chief when it was all your inspiration and hard work, or if I fail to mention it entirely. Believe me, I've been there. By way of excuse I can only repeat the earlier statistic: over 200,000 publications on sex every year. Also, in areas of research where I'm not familiar with the participants, I have had to deduce who were the prime movers from the sequence of authors of a publication, which is not always a reliable indicator.

Scientific writing can be daunting on account of the jargon. The brain is the worst: researchers are constantly dividing this organ into ever-smaller parts with distinct connections and functions, so we end up with names like "the ventrolateral division of the ventromedial nucleus of the hypothalamus"—a tiny neuronal group whose role in sexual behavior I discuss in chapter 3. I've done my best to avoid technical terms, but short of trivializing the whole subject, I've been obliged to include quite a few. I define these when I first mention them as well as in the glossary at the back of the book.

I am most grateful to the researchers who took time to discuss their work with me while I was researching the book, to Todd

Nickerson for his willingness to bare his soul to me on a sensitive topic, to the prepublication reviewers who provided valuable comments and suggestions, and to my editor at Columbia University Press, Miranda Martin, who accepted my proposal and guided it through to completion.

ATTRACTION, LOVE, SEX

1

WHY HAVE SEX?

Why have sex? A dumb question, perhaps. Maybe you've already come up with an obvious answer. But when we probe a little deeper, it turns out to be one of the trickier questions that science has grappled with.

Let's start by asking people for their own thoughts on the matter or, rather, by reviewing a couple of studies that have done this for us. What we'll learn is that there are two hundred and thirty-seven reasons why people have sex. Or, alternatively, just one big reason.

WHAT PEOPLE TELL US

That figure of 237 reasons comes from a research paper published a few years ago in the *Archives of Sexual Behavior*.[1] The paper was titled, plainly enough, "Why Humans Have Sex." It was written by Cindy Meston and David Buss of the University of Texas at Austin. Meston is a clinical psychologist and sex researcher; Buss is an evolutionary psychologist. Both are highly regarded in their fields.

To gather their data, Meston and Buss asked their psychology students to "list all the reasons you can think of why you, or someone you have known, has engaged in sexual intercourse in the past." The most frequent reason given by both men and women was "I was attracted to the person." The least frequent reason given by women was "I wanted to give someone else a sexually transmitted disease." The least frequent reason given by men was "The person offered to give me drugs for doing it."

Between these two extremes the list featured a wide spectrum of reasons, from the mundane to the exotic: "I was curious about sex"; "The person was too hot to resist"; "I wanted to get out of doing something"; "I wanted to change the topic of conversation"; "It was an initiation rite"; "I wanted to get closer to God"; or "It just happened."

Notably absent from the list was any reason connected with procreation. No "I wanted to get pregnant." No "We wanted to have a baby." No "It was time to start a family." The only reason that may have had some connection with procreation was number 152: "I was married and you're supposed to"—but if that was a reference to childbearing, it was an oblique one at best.

Meston and Buss did a good job with this study, in my opinion. Still, it did have some obvious limitations. How well chosen were the participants, for example, for a study that purported to tell us why "humans" have sex? All University of Texas psychology students are humans, for sure, but not all humans are psychology students. What about those who are not?

Let's take a 7,600-mile hop southeastward from Austin. This brings us to the rainforest of Africa's Congo basin, home of the Aka, a pygmy hunter-gatherer people. The lives of the Aka have been studied over many years by the anthropologists Barry and Bonnie Hewlett of Washington State University.[2] Here's what the Hewletts report regarding the Aka people's sex lives.

Couples engage in sexual intercourse about three nights of every week, and three to four times during each of those active nights. They continue to have sex at about this rate throughout their fertile years.

The Aka couples might have sex for the same reasons that Texas psychology students do, of course. They might find their partners "too hot to resist," for example. But that isn't the reason they gave the Hewletts. Rather, both women and men described having frequent sex as a chore—as "hard work" necessitated by their desire to have as many babies as possible. Some Akas did concede that sex could be pleasurable, but pleasure was not their stated motive for engaging in it.

The reason for the Akas' desire for many children was twofold: the high rates of infant and child mortality in the Aka population and the economic and social advantages conferred by having a large family. "I am now doing it five times a night to search for a child," one young Aka man told the Hewletts. "If I do not do it five times my wife will not be happy because she wants children quickly." To keep up this level of performance, Aka men chew on the bark of a certain tree they call *columba*. This may be *Pausinystalia yohimbe*, a tree in the coffee family whose bark is the source of the sexual stimulant yohimbine.

The Aka believe in a concept called "seminal nurture"—the idea that a pregnant woman's fetus requires an ongoing supply of semen for it to survive and grow. This belief is incorrect. (As a biologist rather than a cultural anthropologist, I am permitted to say this.) In fact, if feeding the fetus were the point of the exercise, oral sex would be marginally more useful than vaginal sex, because a man's ejaculate does offer a few calories.

Although seminal nurture doesn't help a fetus grow, frequent sex does increase the likelihood that a woman will become pregnant, for several reasons. Most acts of intercourse are mistimed with respect to the woman's date of ovulation; those that are well timed may not lead to fertilization; and many fertilized eggs die before they ever implant in the woman's uterus. Thus the Aka's belief in seminal nurture, mistaken or not, may be useful in the context of a need to have many children.

Along the same lines, the Aka seem not to practice—or even be aware of—nonprocreative sexual activities such as masturbation or sex between two men or two women. With regard to masturbation, the Hewletts relate the experience of a medical anthropologist, Robert Bailey of the University of Illinois at Chicago, who wanted to obtain semen samples from men of another pygmy people. Bailey had to give the men detailed instructions on how to masturbate, and although the men did come back with semen samples, all but one of the samples contained vaginal secretions mixed in with the semen. All in all, the peoples of the Congo rainforest have sex to make babies, and they don't seem to know of any other reason.

How can two different groups of people—American psychology students and African hunter-foragers—come up with such different motives for having sex? Part of the answer, of course, is that the Americans were college students. Making babies doesn't rank high on most students' to-do lists. It probably ranks higher on their *not-to-do* lists. Yet they are not alone in this: according to an analysis by the Guttmacher Institute, the average American woman spends thirty-one years of her life avoiding pregnancy, something that would be unthinkable to Aka women.[3]

Furthermore, a question starting with "Please list all the reasons you can think of . . . ," as Meston and Buss's question did may be interpreted more as a challenge to creativity than a probe of the participants' actual motivations. It reads like one of those open-ended IQ questions, like "List all the uses you can think of for a copy of *Psychology Today*," in which you get credit for "swatting flies," "starting campfires," and so on. That's very different from how the Hewletts probed the sexual motivation of the Aka men and women.

Meston and Buss's study was not really about why humans have sex; it was about *what people say* when they are asked why they have sex. But by and large, people don't know why they do the things they do, and so they can't tell you. That doesn't stop them from giving you *some* answer, though, because the human mind is a wonderful confabulator.

We need to dig deeper if we want to understand why humans have sex. One direction to dig is deep into the brain so that we can inspect the inner mechanisms that underlie sexual desire and behavior. That's something I'll attempt in later chapters. For now, I want to dig in a different direction—deep into the past, into the question, "Why has sex evolved?" Or, "Given that it has evolved, why doesn't it disappear?"

WHAT MUD SNAILS TELL US

It's natural for us humans to think that sex is necessary because if everyone gave up on sex, our species would quickly go extinct. But why shouldn't we reproduce by splitting in two, like amoebas? If that's too far-fetched a notion, why shouldn't we all be female and reproduce by virgin birth—or PARTHENOGENESIS,[4] as it's called—as some lizards do? Why do we reproduce by sex, a messy, time-wasting, and disease-spreading activity that, among other problems, requires the invention of a superfluous gender—males?

It's easier to think of reasons why we shouldn't reproduce sexually than reasons why we should. Some simple arithmetic makes this clear. Think about a female who reproduces sexually.

Let's say she's capable of having four offspring—two daughters and two sons. The next generation will consist of the offspring of the two daughters, that is, eight individuals in total. The next generation will consist of sixteen individuals; the next, thirty-two; and so on. The numbers will double with each generation until their multiplication is limited by external factors such as a limited food supply.

That sounds efficient, until you compare it with the situation for a female who reproduces asexually, like those lizards. This female's four offspring will be like herself, so all will be asexually reproducing females. The next generation will consist of sixteen females; the next, sixty-four; and so on—the numbers quadruple with each generation. Thus if the two populations are sharing the same environment, the asexually reproducing individuals will rapidly outnumber those that reproduce sexually. Other things being equal, they will monopolize the available resources, causing the sexually reproducing lineage to die out. This twofold cost of sexual reproduction, caused by the necessity to produce males, was initially pointed out by the British evolutionary biologist John Maynard Smith in the 1970s.[5]

Maynard Smith's paper was a theoretical exercise, but it has since been observed in real life. Not in humans, of course, but in a species of tiny snail that lives in the lakes and streams of New Zealand, the New Zealand mud snail, or *Potamopyrgus antipodarum*. What's interesting about these snails is that the females come in two kinds. One kind reproduces sexually, that is, it mates with males; the other dispenses with males and reproduces asexually. Over the last several decades, the snails have been studied by a group led by the evolutionary biologist Curtis Lively of Indiana University. They studied the snails in Lake Alexandrina, which is on the South Island near . . . near nowhere in particular, which makes it the perfect site for field research.

Between 2012 and 2016, Lively, along with his wife, Lynda Delph—who is also a biology professor at Indiana University—and a graduate student, Amanda Gibson, took the long trip to Lake Alexandrina every southern summer, thus skipping a big chunk of the Indiana winter. (*P. antipodarum* is a globally invasive species now found, for example, in Lake Michigan, just a few hours' drive from Indiana University. Presumably it's more than just the weather that necessitates the trip to New Zealand.)

Gibson, who is now an assistant professor at the University of Virginia, told me that the protocol was as follows. Lively donned a wetsuit and swam around with a fine-mesh net, sweeping it through the shallow-water vegetation. (It's somewhat unusual, and laudable, that a senior professor had time and inclination to do this.) Gibson stood in the shallows and received the full nets, whose contents she rinsed to get rid of the large debris. She then handed the snails to Delph, who labeled and packaged them for conveyance to the Edward Percival Field Station, located north of Christchurch. There, all three researchers separated out the tiny juvenile snails—a mix of sexual and asexual individuals—and put them in outdoor tanks, about eight hundred to a tank. They then went home to Indiana, leaving the snails to fend for themselves.

By the time Lively's group returned to New Zealand the following summer, the previous year's juvenile snails had matured into adults and produced offspring of their own. Gibson took samples of the adult females and their offspring back to Lively's lab at Indiana University. To determine the proportions of sexual versus asexual animals, she took advantage of the fact that the asexual snails possess three sets of chromosomes in each cell nucleus, whereas the sexual snails possess only two sets—the normal number in most animals—so the asexual snails have more DNA in each cell nucleus than the sexual snails. She ran hundreds of cell nuclei from each animal through a device called a "flow cytometer," which measures the amount of DNA in each nucleus as it is carried past a sensor in a fast-running stream of fluid.

Gibson found that the asexual females had indeed produced relatively more offspring than the sexual females, and this was true in all four years that she repeated the experiment.[6] In fact, the relative numbers of sexually and asexually reproducing snails corresponded closely to what one would expect based on Maynard Smith's theoretical argument. This study showed for the first time that the cost of sex—more specifically, the cost of producing males—is real and not just the product of a mathematical exercise. In other words, asexual females do have a major advantage over sexual females, at least in *P. antipodarum*.

ONE IDEA: THE RED QUEEN

Nevertheless, sexual snails haven't disappeared, nor have sexual humans or any other of the multitude of species—over 99 percent of all animals and plants—in which sex is either the necessary or an optional means of reproduction. Therefore there must be something about sex that offers an advantage over virgin birth—an advantage great enough to compensate for Maynard Smith's twofold cost of producing males—not just that twofold cost, in fact, but also the substantial costs of finding mates, persuading them to have sex with you, actually getting it on, and contracting whatever diseases your partners may have picked up in the course of their earlier conjugations.

The key feature of sexual reproduction is that individuals inherit a mixture of genes from their parents, half from one and half from the other. Which genes come from which parent is a matter of chance; each sibling inherits a different mix, unless they are identical twins.

What's the advantage of mixing genes? Many theories have been put forward, but there are two main ideas, both of which are likely to be part of the answer. The first is that when genes are mixed, they sometimes end up in a combination that provides a defense against a novel external threat. The threat could be of any kind, but the one that has been most widely discussed is infection by microorganisms or parasites. These organisms are capable of rapid change. When they change, combinations of genes that have previously protected animals against infection may suddenly lose their efficacy. But the shuffling of genes during the process of sexual reproduction means that some individuals in the next generation may have combinations that restore protection. Only a few individuals will get these useful combinations, but those are the ones that survive and reproduce.

This scenario is a never-ending process: the pathogens change their attacks; the animals change their defenses; the pathogens change again; and so on. It's a running battle just to remain as successful as you were before. For that reason, this explanation for sexual reproduction has been called the "Red Queen model," after the Red Queen in Lewis Carroll's *Through the Looking-Glass*, who told Alice that "it takes all the running you can do, to stay in the same place."[7]

The Red Queen model may explain why the sexual snails studied by Gibson are not crowded out of existence by the asexual, parthenogenetic snails, even though the latter can multiply much more rapidly. In Lake Alexandrina, snails risk being parasitized by a certain flatworm, whose eggs the snails may inadvertently ingest. The eggs hatch inside the snail, and the resulting larvae, like aliens in a horror movie, devour every part of the snail that can be eaten without actually killing it. These nonvital parts include the genitals, so the infected snails lose the ability to reproduce. In her experiments Gibson took care to use only uninfected snails, and in their absence, the cost of sex—slower propagation—was evident. In the natural situation, however, this cost may be balanced by the advantage of resistance against the flatworms.

Some further observations on the mud snails support the Red Queen model. Lively's group found that the sexually reproducing females were most common in parts of Lake Alexandrina where ingesting flatworm eggs was more likely to occur. In parts of the lake where there were few eggs, asexual females were more common. This suggested that ever-changing attacks by the parasites promoted the form of reproduction in which new, protective gene combinations could be rapidly assembled. What's more, Lively and colleagues found that over many seasons, the proportions of asexual and sexual snails and the proportions of infected and uninfected snails in these two populations rose and fell periodically in a manner predicted by the Red Queen model.[8]

ANOTHER IDEA: RUBIES IN THE RUBBISH

The other idea about why sex is so popular has to do with MUTATIONS—changes to an organism's genome that may affect its ability to survive and reproduce. This model proposes that sexual reproduction exists because it helps get rid of harmful mutations and preserve beneficial ones.

A newborn human child inherits an average of about seventy new mutations—mutations that its parents did not themselves inherit. This figure was obtained in an Icelandic study that compared the entire genomes of parents and their children.[9]

The mutations crop up in the parents' germ lines—the lineages of cells that develop into sperm or ova. Most mutations occur in the process of cell division. It takes far more rounds of cell division to produce sperm than to produce ova, because a man must generate billions of sperm over his lifetime, whereas a woman creates only a million or so ova, and these are already present (in an immature state) when a girl is born. Thus the great majority of new mutations are inherited from a child's father, and the older the father is, the more new mutations he is likely to pass on.

Many mutations have little or no effect, so the child of an older father may not be impaired in any way. Of those mutations that do have an effect, however, the great majority are harmful, and only a small number are beneficial. That's hardly surprising. You seldom improve computer software by introducing random changes, and the same is true for the genome.

When a species reproduces asexually, natural selection can preserve beneficial mutations, but harmful mutations inevitably come along for the ride—they are unwelcome hitchhikers on the road to the future. And when beneficial mutations arise, it's not likely that they will all arise in the same line of animals, so they can't be combined. Instead, they may end up competing with each other, which often causes one of them to be lost.

When a species reproduces sexually, on the other hand, the random mixing of genes causes the genomes of the offspring to differ from each other. Thus the relative numbers of harmful and beneficial mutations vary from individual to individual; some have more than their share of harmful mutations and some have more than their share of beneficial ones. This potential for sexual gene mixing to separate good from bad mutations increases the power of natural selection to work its magic: individuals with mostly good mutations reproduce more while those with mostly

bad mutations reproduce less, so the genomes with good mutations come to predominate in the population.

According to this model, then, the important difference
between asexual and sexual reproduction is this: with asexual
reproduction, natural selection "sees" the entire genome as a unit;
with sexual reproduction, it "sees" individual mutations, preserv-
ing the good ones and removing the bad ones. This advantage for
sexual reproduction has been called the "Ruby in the Rubbish"
model—the phrase was invented by the British evolutionary
biologist Joel Peck.[10]

WHAT YEAST TELLS US

An incisive test of the Rubies in the Rubbish model was published
in 2016 by a team at Harvard University led by the evolutionary
biologist Michael McDonald.[11] (McDonald is now at Monash
University in Melbourne, Australia.) This group studied brewer's
yeast—a unicellular organism named *Saccharomyces cerevisiae* that
has become a workhorse of not only brewers but also molecular
biologists. The yeast can be induced to reproduce either sexually
or asexually by varying the conditions in which it is grown. The
researchers maintained some samples of yeast cells in conditions in
which they reproduced asexually and others in which they reproduced sexually. Each sample consisted initially of a clone—a set of
genetically identical cells. After the cells had been allowed to reproduce for about one thousand generations, the researchers tested the
"fitness" of the sexual and asexual cells by running the cells in head-
to-head competition with the ancestral strain from which they
were derived. They found that both kinds of cells had increased in
fitness (they outpropagated their ancestors), but the sexual cells had
done so to a much greater degree than the asexual cells.

That result was expected based on previous work. But McDonald's group added a new level of analysis to the experiment—an analysis made possible by the precipitous drop in the cost of DNA sequencing in recent years. They sequenced the entire genomes of both the sexual and asexual populations at intervals of ninety generations over the entire thousand-generation experiment. By this method they identified the mutations as they occurred and followed them over time. The researchers then assayed the harmfulness or benefit of each mutation by transferring the mutation into the ancestral line of yeast and competing this modified line against the unmodified ancestral line. If the new line grew better than the ancestral line, the mutation was a beneficial one; if it grew less well, the mutation was harmful. By these means McDonald and his team were able to reconstruct the history of all the mutations as they appeared and survived or disappeared over the course of the experiment. It was rather like charting the life histories of all the characters in an especially convoluted film noir.

What were the findings? In the asexual population many harmful mutations persisted up to the end of the experiment, by virtue of their unbreakable connections to beneficial mutations. Meanwhile, some beneficial mutations died out in the face of competition from different beneficial mutations in other lines within the population. In the sexual population, by contrast, the mixing of genes allowed the good and bad mutations to separate into different lines; the cell lines with bad mutations were reduced in number or eliminated while those with good mutations persisted. In fact, many of the beneficial mutations vanquished all their competitors, meaning that every surviving cell in the population carried the mutation.

McDonald's findings certainly don't disprove the Red Queen model. What they do show is that the Ruby in the Rubbish model works as theory had predicted—in yeast, anyway. The Red Queen model probably also works in some situations, especially when novel environmental threats are frequently encountered.

What about more complex creatures, such as humans? The relevance of the Ruby in the Rubbish model is made clear by looking at the human genome—more specifically, the Y chromosome. This chromosome is unique in the human genome in that it is not paired with a homologous partner. Most chromosomes come in pairs, one inherited from the mother and the other from the father, and gene exchange between the members of each pair is what makes the separation of harmful and beneficial mutations possible. But the Y chromosome exists as a singleton and then only in males.

Because of the Y chromosome's unpartnered status, the Ruby in the Rubbish mechanism doesn't operate, or at least not in the usual fashion. As a result, most of the Y chromosome's genes have accumulated harmful mutations to the point that they are no longer functional. Many others have disappeared completely, so that the Y chromosome is tiny in comparison with the X chromosome and most other chromosomes. Only a few genes remain fully intact, including an important gene that causes its owner to develop as a male, as well as some other genes involved in male development. For the most part, though, the Y chromosome is a scrapyard of derelict genes, along with massive amounts of "junk DNA," much of which may have originated in ancient viral infections. Most of the rubies have been lost, and the rubbish has spread like an invasive weed. This illustrates the importance of gene exchange over evolutionary time.

The human Y chromosome might have disappeared completely—as has actually happened in some rodents—except that it has invented a devious-seeming trick. As reported by a team led by David Page of MIT, the Y chromosome contains some long palindromic DNA sequences—the sequence runs in one direction at one location on the chromosome and in the reverse direction at another location.[12] Thus the genes located within these sequences, which include all the genes that are still functional, exist in two copies that are mirror images of each other. This allows the chromosome, by folding back on itself, to exchange copies of the same genes that are located within the palindromic sequences. It's as if you corrected the typo in "Ma*r*am, I'm Adam" by folding the phrase back on itself and using the "d" to correct the faulty "r." This trick could almost be considered a form of autoerotic sex. It allows the Rubies in the Rubbish mechanism to preserve the few Y chromosome genes that are still functional.

WHAT LIZARDS TELL US

What about those species, like the parthenogenetic lizards, that never reproduce sexually? Does their existence show that sex isn't necessary at all?

The best-known parthenogenetic lizard, the desert grassland whiptail, or *Cnemidophorus uniparens*, inhabits dry scrublands of the American Southwest. (It's also known as *Aspidoscelis uniparens*.) The species arose by hybridization between two species of sexually reproducing lizards, which still exist in the same areas. The asexual lizards retain some behavioral memory of their sexual ancestors, in that they still go through the motions of sex. This behavior—a coupling between two parthenogenetic females—has been given the unromantic name of "pseudocopulation."[13] Although no sperm or ova are exchanged, pseudocopulation does have a useful function: it promotes the maturation of eggs through a hormonal mechanism.

On the face of it, the existence of *C. uniparens* challenges both explanations for sexual reproduction that I described earlier. These lizards can't use sex to assemble beneficial combinations of genes, nor can they use it to get rid of harmful mutations. So why don't they die out?

The truth is that they may die out over long periods because of their inability to reproduce sexually. Even so, they are replaced by new lines created in new hybridization events. In fact, researchers associated with the Howard Hughes Medical Institute have succeeded in replicating this process in the lab. By mating males and females of the two ancestral species, they produced asexual females that continued to reproduce parthenogenetically.[14]

No one has observed the appearance of new lines of whiptail lizards outside the laboratory. At least one such natural event has been observed in another animal, however: the marbled crayfish. This new species of asexual crayfish arose in Germany in the early 1990s as a result of a single mutation in a sexual species of crayfish.[15] The marbled crayfish now number in the billions; they threaten to displace other crayfish species in many parts of the world, thanks in part to their asexuality. They are genetically identical to one another, however, so that when the right parasite appears, it may wipe out every last one of them, putting an abrupt end to the species' meteoric career.

In short, the asexual lizards and other asexual animals, fascinating though they are, fail to invalidate existing hypotheses about sexual reproduction. They can do without sex, but probably not forever.

WHAT BDELLOID ROTIFERS TELL US

Before we conclude that sex is necessary, however, I have to acknowledge that there is one kind of animal that *can* do without sex forever. These are microscopic invertebrates called BDELLOID ROTIFERS. That's DELL-oid, which means "leechlike"—they look and act like tiny versions of leeches—and ROTE-ifer, which means "wheel-bearing," a bdelloid rotifer sports a circlet of ever-whirring bristles around its mouth, which doubles as its anus. The Swiss microbiologist Bernard Jenni has posted an entertaining video of them on YouTube.[16]

Bdelloid rotifers are strange animals in ways other than their looks. One of their many talents is that they are more resistant to ionizing radiation than any other animals. They also survive complete desiccation. Once dry, they wait out the drought—for years if necessary—until they encounter water again, whereupon they reassume their normal appearance and carry on as if nothing had happened. A bdelloid rotifer might survive a trip from Earth to Mars on the outside of a spacecraft.

Like parthenogenetic lizards, bdelloids are all female. They haven't had sex for millions of years, according to molecular-genetic studies by a team led by Matt Meselson of Harvard University.[17] Like the lizards, they are often cited as a challenge to the idea that sex is necessary.

It turns out, though, that bdelloid rotifers have a trick up their tiny sleeves: they can take up DNA from the environment and patch it into their own genomes. This happens during the episodes of desiccation, when the cell membranes become leaky (allowing the foreign DNA to enter) and the rotifers' own DNA is subject to random breakage (allowing the foreign DNA to be spliced into the breaks). The foreign DNA can come from any source—other rotifers, bacteria, fungi, or even plants. Although the DNA can come from anywhere, the bdelloids preferentially retain DNA from species related to themselves, species whose genes might be useful to them.[18]

It is this "horizontal gene transfer" that likely takes the place of sexual reproduction. The process harks back to the bacterial world; it is one method by which bacteria acquire genes confer-ring resistance to antibiotics. In a sense, bdelloid rotifers do have sex, but in a random, promiscuous fashion that doesn't involve direct contact between organisms.

SEX BEYOND REPRODUCTION

But there's one aspect of sexual reproduction that has opened the door to all kinds of strange-seeming deviations—and I don't mean that in a bad way.

The aspect I'm referring to is this: sexual reproduction requires complicated behaviors before and after pregnancy, especially in mammals. Before pregnancy, there's courtship and sexual inter-course; after pregnancy, there's nursing and rearing, which may continue for years. But sexual reproduction doesn't require any behaviors *during* pregnancy. Some women have not even known they were pregnant until the day they gave birth,[19] and there's at least one documented instance of a woman being in a coma for the entire duration of her pregnancy.[20]

Nature has to provide the motivation for sexual intercourse, but she doesn't have to provide the motivation for pregnancy, because pregnancy looks after itself. Once a baby is born, a new set of motivations, falling under the general title of parent-child bonding, comes into play.

Perhaps Nature should have thought this through more care-fully because by offering all kinds of carrots before pregnancy—sexual attraction, arousal, the pleasure of sexual contact, and the reward of orgasm—while offering no carrots at all and perhaps even some sticks for the duration of pregnancy, it has invited animals with any degree of creativity to grab the carrots and dodge the sticks.

We thus see the emergence of numerous nonreproductive sexual behaviors of the kind that St. Thomas Aquinas, the thirteenth-century arbiter of sexual ethics, condemned so fervently. Noncoital sex acts (e.g., oral or anal sex), masturbation, same-sex behaviors, contraception, sex during menstruation, and sexual kinks—if Aquinas knew about them—were all sinful in his eyes. Those 237 reasons to have sex listed by Meston and Buss's students? The only one that Aquinas might have let pass was the one mentioned earlier: number 152: "I was married and you're supposed to," and then only if contraception wasn't used. The sexual motivations of the Aka, on the other hand—sexual pleasure subordinated to the rational purpose of procreation—would have earned Aquinas's unqualified praise, if he had known of the Aka.[21]

In terms of their evolutionary origins, the motivations for nonreproductive sexual behaviors are accidental by-products of the motivations for reproductive behaviors. That doesn't mean that they are maladaptive, however—that is, they don't necessarily hinder the perpetuation of the genes that generate these motivations. In fact, they can be adaptive.

That becomes clear if we consider our oversexed primate relatives, the bonobos.[22] A female bonobo is sexually active not just when she is ovulating, as many other mammals are, but for her entire sixty-day menstrual cycle, with the exception of a few days around the time of menstruation. She is also sexually active while breastfeeding an infant, even though breastfeeding suppresses ovulation. In other words, female bonobos, like women, may initiate or respond to sexual advances even when they are incapable of becoming pregnant.

In addition, bonobos engage in sexual behaviors that by their nature are nonreproductive, such as same-sex contacts and contacts between adults and sexually immature juveniles. Regarding same-sex contacts, two males may rub their penises together while hanging upside down from the branch of a tree. Female pairs engage in a behavior called "genito-genital rubbing"— another unromantic technical term—in which they sweep their vulvas against each other with a rhythmic side-to-side motion. There are no exclusively "lesbian" bonobos, as far as we know, but all females have sexual contacts with other females more frequently than they do with males.

Bonobos evidently engage in these nonreproductive behaviors because they are pleasurable. But in doing so, they are not simply cheating sex of its reproductive goal. Rather, they have found an adaptive value for the behaviors. According to the primatologist Frans de Waal of Emory University, the behaviors help resolve conflicts and promote social cooperation. Sex between females strengthens female alliances, and these alliances help females achieve a measure of social dominance over the physically stronger males.

These behavioral adaptions must have a biological basis. For females to be sexually active when they are not ovulating has required evolutionary changes in their neuroendocrine control systems. Also, it is an anatomical trait that makes genito-genital rubbing possible—the forward-facing position of bonobos' vulvas. There must also be changes in brain circuitry that make same-sex individuals into attractive sex partners, although these changes have not been studied. In other words, even though nonreproductive sex wasn't what Nature originally intended, she has acknowledged the inevitable and made the best of it.

CONCLUSIONS

I've explained why the evolution and persistence of sexual reproduction is paradoxical: *asexual* reproduction should be twice as efficient at producing offspring. I've discussed two main ideas for why, in spite of this disadvantage, the capacity for sexual reproduction remains universal or nearly so. One idea is that the central feature of sexual reproduction—mixing genomes—helps protect against environmental threats such as parasites (the Red Queen model). The other is that mixing genomes helps get rid of harmful mutations and combine beneficial ones (the Rubies in the Rubbish model).

Given that there's evidence supporting both models, it's reasonable to conclude that each represents part of the reason why sexual reproduction has been so successful. Yet the aesthete in me rebels against that conclusion. To my mind, there should be a single, beautiful reason for something as paradoxical—and yet so central to life on Earth—as sex.

If I had to choose between the competing models, I'd have no hesitation: the Rubies in the Rubbish model is more beautiful. Parasites come and go, but mutations lie at the core of why sustaining life across the generations is so challenging. It may be that the Red Queen model can be thought of a feature or subroutine of the Rubies in the Rubbish model: without her Rubies, after all, the Red Queen would eventually stop running and be swept away.

The remainder of this book focuses on sex in our own species. Humans, or the Westernized humans who have been the focus of most research, have developed the nonreproductive functions of sex to an extraordinary degree—so much so that Meston and Buss were able to list 237 reasons for having sex without once mentioning reproduction. Yet we can't escape our evolutionary history: sexual attraction, the topic of the next chapter, is all about making babies, even if we never have any.

2

ATTRACTION

S exual attraction happens when one person is attractive and another is attracted, but these two aspects of attraction entangle each other in strange ways. Beauty is an objective attribute of an attractive person, but it's also a subjective assessment by the person who is attracted. Sorting out these two sides of attraction has occupied legions of psychologists. Their findings are sometimes confusing, sometimes contentious, and, once in a while, revelatory.

SKINNY OR FAT?

Surely your concept of the ideal female form doesn't change simply because you've just enjoyed a good meal? But it does—if you're a male college student, at least. That's the conclusion of a study by the British psychologists Viren Swami and Martin Tovee.[1] Beauty, according to their findings, lies not in the eye the beholder, but in his stomach.

Swami and Tovee waylaid young men as they entered or exited a college dining hall at dinner time. They asked the men to peruse a booklet containing photographs of fifty women with their heads obscured. The women's BODY MASS INDEX (BMI) had been measured. (According to the usual criteria, a BMI below 15 is "emaciated," 15 to 18.5 is "underweight," 18.5 to 24.9 is "normal," 25 to 29.9 is "overweight," and above 30 is "obese.") The men were asked to rate the women's attractiveness, as well as their own level of hunger. The researchers did not ask the men about their sexual orientation. Assuming that the great majority were heterosexual, we may suppose that they interpreted "attractive" to mean "I'd like to have sex with her."

There was a striking difference between the judgments by the hungry and the full men. As figure 2.1 shows, the full men preferred women with BMIs around 19—that is, close to the low end of the normal range. The hungry men liked women with BMIs through the entire normal range and well into the overweight range. In fact, although attractiveness declined with increasing weight for both groups, all BMIs above normal were

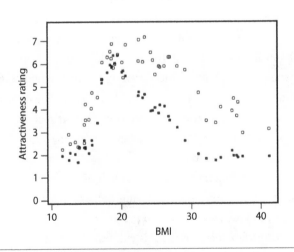

FIGURE 2.1. Hunger and attractiveness.

Source: From V. Swami and M. J. Tovee, "Does Hunger Influence Judgments of Female Physical Attractiveness?" *British Journal of Psychology* 97 (2006): 353–63.

more attractive to hungry that to satiated men. Underweight BMIs, on the other hand, were about equally unappealing to satiated and hungry men. Swami and Tovee's study replicated and extended the results of an earlier study by two American researchers, Leif Nelson and Evan Morrison.[2]

My first thought when I saw these data was that hungry men view women as edible, which would obviously bias them toward plus-size women. Swami and Tovee don't share that ghoulish interpretation, however. Rather, they place their findings in the context of several other studies demonstrating the effects of situational factors on what men find attractive in women.

It's well known, for example, that men prefer fatter women in countries experiencing food shortage. In fact, some African countries, such as Nigeria, had a traditional custom involving what was called the "fattening room."[3] A woman seeking a husband was sequestered in this room, where she was fed a costly, rich diet with the aim of increasing her weight. After several weeks or months of this regimen, she and her competitors paraded before eligible young men, who chose their brides largely on the basis of their BMI—the heavier, the better. Parents who could not afford the fattening regimen wrapped their daughters' midriffs in many layers of cloth to simulate fatness.

The usual interpretation of this custom is that men in a state of food insecurity are more attracted to women who, by their ample girth, demonstrate superior access to resources. With Westernization and increased food availability, however, the tradition of the fattening room is dying out: women are rebelling against it, and men are coming to prefer slimmer brides.

Tovee's group documented this cultural shift in a clever way.[4] Using a method similar to the one they employed with the British students they first measured the BMI preferences of Zulu men living in their South African homeland. (The data were collected by Roshila Mangalparsad of the University of Natal.) These men rated all women with BMIs above about 20 as equally attractive—even women who, by Western medical standards, would be diagnosed as severely obese. They then tested Zulu men who had moved to the United Kingdom. Within eighteen months of living in Britain, these men judged obese women to be significantly less attractive than women with BMIs in the normal range. And African men who had grown up in the UK had the same preference for women with low-normal BMIs, and the same dislike of obesity, as was shown by white Britons.

So some cultural or environmental factor has a strong and malleable influence on how men respond to women's body shapes. But what is that factor? It could be food availability, as suggested by the study of the British college students. In other words, the Zulu men in their homeland might have been food-deprived, either over their lifetimes or on the particular day when their preferences were assessed. Alternatively, it might be a general influence (or lack of influence) of Western culture.

To sort out these possibilities, Tovee and seven colleagues turned their attention to Nicaragua.[5] They measured the BMI preferences of both men and women in three locations: the capital, Managua, where people have long had access to television and frequently watch Western-style soap operas; a village in which the inhabitants watched lesser amounts of television; and another village, in which people watched no television at all because there was no electricity service. The more TV people watched, the more they adopted Western standards of beauty, rating women of low-normal BMI as the most attractive. This was true for both male and female participants. The participants' state of hunger at the time of testing—estimated by how much time had passed since their last meal—also influenced their judgments, in the same way as it did in the case of the British students.

Many of the Nicaraguan women who watched a lot of television had placed themselves on diets with the aim of aligning their bodies with their new ideal of beauty. Similar responses have been noted in several developing countries. In Fiji, for example, Anne Becker of Harvard University observed an outbreak of anorexia and other eating disorders among young women soon after they were exposed to television and American soap operas.[6]

In all of Tovee's studies, as well as those of other researchers, the low-rated attractiveness of emaciated or underweight women was a constant, regardless of the participants' state of hunger or exposure to Western culture. This is understandable, perhaps, because significantly underweight women are less able to carry a pregnancy to term than are women of normal weight. In fact, they may not ovulate or menstruate at all, in which case they won't be able to become pregnant in the first place. Thus men's relative lack of attraction to very underweight women could well be an evolutionary adaptation, one that increases men's reproductive success by focusing their attention on fertile women.

Although men in food-secure cultures such as our own don't find significantly underweight or emaciated women attractive, the BMIs that are rated most attractive—18 or 19—are at the lower end of the normal range or even extend slightly into the underweight range. According to analyses by the anthropologists William Lassek and Steven Gaulin of the University of California, Santa Barbara, American women in the most attractive range of BMIs are significantly less fertile and probably also less healthy than women with higher BMIs.[7] It may be that men underestimate the age of slender women. Given that younger women are generally healthier and more fertile than their elders, the men could be using slimness as a (mistaken) cue to health.

Alternatively, men may not be guided by an evolutionary-psychological (evo-psych) imperative to mate with the healthiest partners. Rather, they may be attracted to whatever body shape is more difficult to achieve and therefore rarer. A BMI under 20 is certainly difficult to achieve when food is plentiful: the average BMI of American women—and American men, too—is 29, which is not just overweight but bordering on obese. A preference for difficult-to-achieve BMIs could therefore explain the perceived attractiveness both of fat women in food-insecure societies and of thin women in affluent Western societies.

So much for men's attraction to women's bodies. But what about women's attraction to men's bodies? This is a simpler story. Women are attracted to men who look strong. A 2017 study led by Aaron Sell of Griffith University in Queensland, Australia, is one of many that illustrate this finding. The stronger that raters judged a man to be (or the stronger the man actually was, based on physical testing), the higher his attractiveness rating.[8] The most important visual cue to bodily strength is the ratio of chest or shoulder width to waist width. That is, a V-shaped torso is the desirable shape, as one might guess from the photographs that populate men's magazines. In 2022, a group led by Linda Lidborg of Durham University in the UK published a META-ANALYSIS of ninety-six prior studies conducted in diverse cultures. Bodily strength or muscularity were the only consistent predictors of men's mating and reproductive success.[9]

This finding screams "evolutionary adaptation." During most of human evolution, strength must have been one of men's most useful physical attributes, just as it is for males of many other species. But that's no longer true, even for fighters: "Combat roles no longer uniformly require sheer size or muscle," asserted a U.S. federal judge in 2019, part of his justification for ruling that the selective service draft should not be limited to men.[10] Today, a strong-looking physique is advantageous in only a few highly specialized occupations, such as nightclub bouncer or gay porno actor. But the mills of evolution grind slowly, and it may take millennia for spindly dweebs to turn women's heads.

In those nonhuman species in which females choose mates on the basis of some physical characteristic, their preference is often open ended. In some avian species, such as widowbirds, females prefer males with long tails. Researchers have kitted out male widowbirds with "tail extenders"; thus accoutered, the birds easily outcompeted their rivals for the favors of females.[11] Similarly, Sell's group found that the attractiveness of strong-looking men has no obvious end point. Thus men who artificially enhance their muscularity with hormones, as Arnold Schwarzenegger did in his prime, outcompete men with naturally toned bodies. Sell's findings run counter to the common belief that women find very muscular men unattractive. Perhaps in real life, women's attraction to such men is counterbalanced by an element of fear.

There's much more to bodily attractiveness than what I've just described. Overall shape of the torso, defined as the waist-to-hip ratio, is relevant. The hourglass shape is generally considered most attractive in women, whereas something closer to a tubular shape is preferred in men. Breast size has been the topic of numerous studies, and women's quest for the most attractive breasts—by enlargement, reduction, or lifting—powers a large industry. But in general, all these studies point to the same conclusion: that judgments of bodily attractiveness are rooted in a mix of evolutionary imperatives, cultural forces, and individual quirks.

WHAT'S IN A FACE?

In the case of judgments about the attractiveness of faces, it's more complicated, and no one understands the complexity better than Lisa DeBruine. Her research group at the University of Glasgow, Scotland, is devoted to faces and how they are perceived. DeBruine's work is socially relevant, because good-looking people don't just attract sex partners; they also enjoy many undeserved advantages in life, as economist Daniel Hamermesh laid out in depressing detail in his 2011 book, *Beauty Pays*.

Perhaps the most important attribute of faces, when it comes to attractiveness, is how male-like or female-like they are. Digital morphing techniques allow this attribute to be represented as a continuum ranging from hyperfeminine—that is, more feminine than any real woman's face—to hypermasculine, with an androgynous face at the midpoint. In general, men find women's faces more attractive the closer they lie to the feminine end of the spectrum.

According to a well-known evo-psych interpretation of this finding, a feminine appearance results from high estrogen levels during puberty and/or in adult life, and these high levels correlate with fertility. The idea, in other words, is that by preferring women with very feminine features, men are boosting their own likely reproductive success. Indeed, some studies with limited numbers of participants have reported a relationship between women's estrogen levels and their facial attractiveness. More recently, though, DeBruine's group conducted a much larger study and found no such relationship, casting the popular evo-psych theory into doubt.[12]

DeBruine has thrown a wrench into an even more widely cited evo-psych idea: the notion that women's judgments of men's faces vary depending on where they are in their menstrual cycles and do so in a fashion that serves their reproductive interests. Starting in the 1990s, several research groups reported that women were attracted to more masculine-looking men during the fertile days of their cycle (at or just before ovulation) than on other days. In a study led by Victor Johnston of New Mexico

State University, for example, the male face that was most attractive to the average ovulating woman moved toward the male end of a 1,200-step male–female continuum by 29 steps, a small but significant shift.[13]

Popular accounts of such findings greatly exaggerated the effect, as if women were drawn to hired assassins while fertile and to amiable provider types at other times. The evo-psych interpretation is that women seek sex with very masculine men for their supposedly high-quality sperm, whereas they offer sex to less masculine men as a means to cement relationships and obtain support in parenting. This was the so-called dual mating strategy hypothesis.

Bolstering that idea was the belief, supported by some early DNA studies, that highly significant numbers of children are not the biological offspring of their mothers' husbands or regular partners; rather, they were fathered by the cable TV repairman or some other tool-belted stranger. That turns out to be largely untrue, however. According to recent studies, no more than about one in one hundred children is fathered by a man other than the mother's husband, and this seems to have been true across most cultures and historical periods.[14] Women are not as devious as previously supposed—or if they are, they take care that no pregnancies ensue. This suggests that women are not pursuing a dual mating strategy or, if they are, not with any great enthusiasm.

DeBruine's group, in a study led by Benedict Jones, attempted to replicate the findings of Victor Johnston and others but failed. Though they recruited much larger numbers of participants and employed more precise techniques to assess the women's fertile or nonfertile status, their judgments of male attractiveness showed no cyclical shifts whatsoever.[15]

The findings of any study may turn out to be incorrect, even one as well-designed as DeBruine's. Still, unless an equally well- or better-designed study confirms the original claims, the hypothesis that women adopt a dual-mating strategy is no longer well supported by the evidence, and it is quite likely incorrect. Whether this revision will percolate into public awareness is questionable, however. Another plausible myth about the menstrual cycle—the belief that women who live together menstruate in synchrony—shows no sign of loosening its hold on the public imagination, even after several failures to replicate the original observations.[16]

DeBruine's findings were not all negative, however. Her group found that a woman's general level of sexual desire does increase near to ovulation, even if there is no change in what she wants to do sexually or with whom. This finding has been reported before, and it is hardly surprising. Females of most mammalian species refuse to engage in sex except when they are capable of becoming pregnant. Only some of our closest relatives, such as chimpanzees and bonobos, are more flexible in that regard, as was mentioned in the previous chapter. It's understandable, therefore, that women's sexuality might retain some trace of sex's original function: making babies.

There's much more to the attractiveness of faces than their masculinity or femininity, of course. Facial symmetry is important; faces are more attractive the more symmetrical they are, and skin quality also plays a role. Another factor is the racial appearance of faces. Early studies suggested that people rate faces of their own race or ethnicity as more attractive than those of other races, as if familiarity was the key to attractiveness.

Studies using computerized face morphing have found something different, however. An Australian-Japanese research group created averaged images of Caucasian and Japanese faces and used them to generate a series of synthetic faces that morphed between the two. The faces rated most attractive lay on the intermediate (so-called mixed-race) part of the spectrum rather than at either of the single-race extremes. This was true whether or not the persons doing the rating were Australian or Japanese.[17] Similar findings have been obtained by American psychologists using European American and African American faces as the starting points: Again, the mixed-race faces were judged most attractive, regardless of who was doing the judging.[18]

This finding appears to hold with unmanipulated real-life photographs. Michael Lewis of Cardiff University in the UK found that his students consistently found photographs of mixed-race individuals more attractive than those of single-race individuals.[19] Lewis suggested that mixed-race individuals are more attractive because of "hybrid vigor"—that is, gene mixing has provided them with some protection against the deleterious effects of inbreeding. But here's a caveat: for the faces in his study, Lewis used photographs from Facebook groups with self-defined titles like "Mixed race and proud of it." To confirm Lewis's findings, the study needs to be repeated with photographs of individuals chosen more randomly from the populations of single- and mixed-race individuals.

SMELLING GOOD

Besides the visual appearance of bodies and faces, other senses come into play, including hearing and smell. Much work has been done on voice quality, and the general findings are similar to those for faces: men's voices are more attractive the more male-sounding (for example, lower-pitched) they are, while for women's voices, it's the opposite. In addition, more subtle characteristics of the voice influence vocal attractiveness—characteristics described by abstruse technical terms like "harmonic-to-noise ratio," "spectral tilt," and "vowel-space dispersion," which provide information about the speaker's sex, youthfulness, and health.[20]

The sense of smell—olfaction—influences attraction but in mysterious and contentious ways. One idea is that people are consciously or unconsciously sensitive to odors controlled by the MAJOR HISTOCOMPATIBILITY COMPLEX or MHC genes. This a set of genes whose protein products help the immune system distinguish self from nonself. (In humans, the MHC genes are also called "human leukocyte antigen," or HLA, genes.) It has long been known that a mouse can tell by olfaction whether another mouse possesses MHC genes similar or dissimilar to its own, and it uses this information to mate preferentially with MHC-dissimilar partners. In this way it's thought that mice reduce the likelihood of inbreeding and maximize diversity among their offspring's MHC genes, to the benefit of their immune systems.

In 1995, Claus Wedekind and his colleagues (then at the University of Bern, Switzerland) conducted an experiment to see if the same was true for humans, their now-famous "sweaty T-shirt study."[21] Heroic volunteers—female students—were tasked with sniffing T-shirts worn by male students over two consecutive nights. Wedekind reported that, like the mice, the women preferred the odor of males whose MHC genes differed from their own. In addition, the women said that the odors of these MHC-dissimilar men reminded them of the odors of their regular partners—and did so significantly more than the odors of the MHC-similar men. Wedekind concluded that women can distinguish MHC-similar from MHC-dissimilar men by olfactory cues and that they make use of this ability in choosing their regular sex partners. He did not test the MHC status of those partners, however, which could have helped nail down his second conclusion.

Wedekind ended his paper with a warning. Individuals who use deodorant may put this whole mechanism out of kilter, he suggested, in which case they risk parenting biologically disadvantaged children. Wedekind could have made other practical points: for example in cultures where marriages are arranged by parents, MHC testing could improve the outcome of the matchmaking process.

That's if Wedekind's findings are correct. Wedekind evidently thinks so, because he later published a study (with Sandra Füri) that replicated the original findings and, for good measure, demonstrated the same effect with the sexes reversed: with men smelling women's T-shirts.[22] But another Swiss group, led by Fabian Probst, who is also at the University of Bern, attempted to replicate this latter part of the study with more subjects and better techniques and failed. Probst found no evidence whatsoever that MHC similarity or dissimilarity affected men's ratings of women's attractiveness.[23]

Wedekind critiqued Probst's study, Probst critiqued Wedekind's critique—and so it goes. Although the matter is not fully resolved, the current evidence suggests that the MHC effect relates to women's choice of male partners and not vice versa. That would be consistent with women's greater choosiness in many aspects of the mating game, a choosiness that can be traced to the huge investment that women must make in childbearing, an investment that they need to make wisely. In any event, more studies are called for to confirm or refute these conclusions, so sniffing semi-rancid T-shirts may become a rite of passage for students at the University of Bern.

In many nonhuman species, PHEROMONES play an important role in mating behavior. Pheromones are chemical signals that influence the behavior of other members of the same species— conspecifics. Some female insects release sex pheromones into the air, for example, thus luring prospective mates from blocks away and triggering an automatic sequence of mating behaviors. Besides these volatile pheromones, other pheromones are transferred from animal to animal by direct contact.

Some mammals and other vertebrates possess a specialized structure within the nose, named the VOMERONASAL ORGAN or VNO. Sensory neurons within the VNO respond to pheromones and other substances, more broadly characterized as CHEMOSIGNALS, that are released by conspecifics. These neurons have AXONS (output fibers), which enter the brain and terminate in regions responsible for reproductive and parenting behaviors. In mice, destruction of the VNO, or the suppression of its function by genetic manipulation, has radical effects on behavior. According to Catherine Dulac and her colleagues at Harvard University, female VNO-impaired mice display a variety of sexual behaviors normally shown only by males, and male VNO-impaired mice lose the ability to distinguish males from females.[24] On the other side of the Charles River, however, Michael Baum and his colleagues at Boston University have reported that some sex pheromones act via the main olfactory mucosa rather than via the vomeronasal organ.[25]

Although it's been claimed that humans also possess a functional VNO, the consensus is that this organ is either absent or vestigial and nonfunctional in our own species. The genes that, in mice, code for the VNO's specialized receptor molecules are also present in humans, but they, too, are vestigial—that is, they have accumulated so many mutations that they no longer code for anything. Thus if human sex pheromones exist, they are probably sensed by the main olfactory mucosa, not the VNO.

But do such pheromones exist? Most attention has focused on two steroids that are generally referred to by their abbreviations

AND and EST. AND is ANDROSTADIENONE, a constituent of men's armpit sweat; EST is ESTRATETRAENOL, which was first isolated from the urine of pregnant women but is also present in other bodily secretions. These two compounds were identified in the 1990s and promoted as human pheromones by the University of Utah physiologist Louis Monti-Bloch and several colleagues; they started a company to sell perfumes and colognes laced with them.

If you imagine that these two substances are come-hither commands written in the language of chemistry, let me disabuse you. They are come-hither *suggestions* at best, and perhaps not even that. Some researchers have reported that these pheromones have beneficial effects on mood, either of the person daubed with them or of others who come within sniffing range of that person. An Israeli study reported that EST has a positive effect on men's social cognition, especially regarding intimacy.[26] A Chinese group reported that AND and EST bias the perception of gender: AND biases women to judge that a gender-ambiguous figure is a man, the group claimed, whereas EST biases men to judge that such a figure is a woman.[27] An Australian group repeated this experiment using gender-ambiguous faces rather than figures and found no effect of either AND or EST, whether on judgments of gender or attractiveness.[28]

In short, the current scientific literature on the effects of AND and EST have provided some suggestive but not dramatic findings. They don't appear to act as sex pheromones in humans, at least in the sense that the term has been applied to other species, because they don't seem to trigger or modify any overt behavior. They may have mild effects on mood and cognition that could influence sexual interactions. I'll have more to say about AND and EST in the chapter on sexual orientation.

BIG DATA

All studies that ask participants to rate attractiveness—whether on the basis of facial photographs, odors, or anything else—have one serious limitation: the participants may not be completely forthright in their answers. In fact, people may not be fully conscious of what turns them on or off.

Psychologists have devised various methods to sidestep this problem. For example, when people are given photos of pairs of faces or bodies, they will spend most of the time looking at the face they find more attractive.[29] Also, people's pupils dilate when they view sexually arousing images.[30] Measuring such phenomena can provide an estimate of attraction and attractiveness that doesn't depend on consciously expressed ratings. Methods of this kind are particularly useful when assessing the preferences of infants or young children who can't yet express their feelings in words. It has been reported, for example, that infants as young as two to three months look longer at faces that adults have judged to be more attractive.[31]

A quite different approach is to study the behavior of adults in real-life situations. Do they walk the talk, in other words? The advent of big data has opened this question to scientific inquiry. In the realm of sex, big data means the statistical analysis of millions of porn searches, dating site messages, sex-related Google entries, and the like. The findings of such studies uncover yawning gaps between what people say and what they actually think and do, as emphasized in the titles of popular books by data analysts: *A Billion Wicked Thoughts What the World's Largest Experiment Reveals About Human Desire* (Odi Ogas and Sai Gaddam, 2011), *Dataclysm: Who We Are When We Think No One's Looking* (Christian Rudder, 2014), and *Everybody Lies Big Data, New Data, and What the Internet Can Tell Us About Who We Really Are*

(Seth Stephens-Davidowitz, 2017). The findings of these data-mining studies are instructive, though not always edifying.

Big data has been particularly informative about the relationship between age and attraction, with regard both to the age of the attractor and that of the attractee. Christian Rudder analyzed data from the dating website OkCupid, which he helped found. On this site, members can message other members in whom they are interested; they can also set lower and upper age limits to control which other members may message them. Judging by the age filters, both women and men are attracted to increasingly older members as they themselves age. Judged by the actual messages they send, however, men are most attracted to women who are substantially younger than themselves. In fact, when simply given faces to rate, men's preferred faces remain stuck at around age twenty or twenty-one, no matter how old the men may be. Women, on the other hand, are most attracted to faces of men of about the same age as themselves, at least until their forties.

These data describe averages, of course, and there are individuals who buck the trends. Most interestingly, a subset of men in their twenties are most attracted to women in the thirty-five to forty-seven age range. This is the much-publicized MILF ("mothers I'd like to fuck") phenomenon. Luckily there is a complementary and equally well-publicized phenomenon: middle-aged women, or "cougars," who seek out much younger men. Every kink has its "antikink," as we'll see in chapter 7.

The analysis of big data also reveals how the criteria for attractiveness, as measured by "honest" data such as porn searches, can change over time. As an example, Seth Stephens-Davidowitz analyzed interest in large butts. Up until around 2010, he reported, this was largely a phenomenon found among Blacks. More recently, searches for "big-butt porn" (mostly by men, in all likelihood) and for methods to enlarge one's own butt (mostly by women, we may assume) have increased dramatically throughout the U.S. population. Stephens-Davidowitz attributed this development to the influence of one celebrity—Kim Kardashian. Her ample posterior has been the focus of more attention than that of any woman since the Venus of Willendorf.

CONCLUSIONS

All in all, research into attractiveness doesn't inspire complete confidence in the scientific approach to sexuality, because there have been so many disagreements and contradictory findings. Regarding this particular field, the so-called replication crisis in psychology is very much in evidence. And I haven't even touched on several topics that are even more contentious than those I've discussed. Is averageness a key criterion for attractiveness, for example? And are faces most attractive if they resemble the viewer, as has been reported in some studies? Even if true, this last idea lacks a ready explanation in terms of evolutionary psychology; prehistoric humans had limited opportunities to view and study their own faces, after all—except for Narcissus, who fell in love with his own reflection in a spring-fed pool.

One among the many solutions to the replication crisis that have been offered is simply to increase the numbers and diversity of participants included in studies. Lidborg's meta-analysis of men's muscularity as a cue to attractiveness did that, and in doing so, it bolstered an evo-psych understanding of sexual attraction. So did the big data studies just mentioned. But big data studies have problems of their own. One is that people who don't search for partners online or don't read English, are often excluded from studies of sexual attraction—Aka tribespeople, for example.

I also haven't touched on the question of how personality and behavior influence attractiveness. Because personality reveals itself over time rather than at first meeting, however, I will cover this topic when I discuss sexual relationships.

Finally, we tend to forget that sexual orientation is all about sexual attraction—to men, women, or both. Because of the vast scientific literature on this topic and its social relevance, I devote an entire chapter to it—chapter 4.

3

AROUSAL

Sometime in the 1960s, a twenty-two-year-old man was referred to neurosurgeons at Johns Hopkins Hospital in Baltimore for the treatment of temporal lobe epilepsy. His history was as follows.

> At the age of five this male patient began having frequent staring spells, which were controlled with phenobarbital. . . . He was free of seizures until age 18, when he suddenly went into a trance while at home with his six-year-old sister. The mother heard him call the girl's name and found that he had stripped her and was attempting intercourse. He regained awareness when his mother gave him a hard slap.
>
> He then began to experience frequent trances characterized by tightening of his body and a brief feeling of sexual climax [orgasm]. These attacks were usually associated with a feeling of fear. His seizures would last about ten to 20 seconds and recurred as often as ten times a day. . . . As far as could be ascertained his ictal [seizure-associated] feelings of sexual climax were not accompanied by erection and certainly not by ejaculation.[1]

Epilepsy arising in the temporal lobe often shows itself by absence attacks or by intense but short-lasting emotional states that have no external cause. In a minority of cases like this one, sexual arousal or inappropriate sexual behavior can occur during or after the seizure or for the first time in patients after they undergo surgical removal of part of the temporal lobe for the relief of epilepsy.

The visible part of the brain's TEMPORAL LOBE consists entirely of cerebral cortex, but buried deep within it are other forebrain structures, including one named the AMYGDALA; the word means "almond" and is a reference to its shape. (There are two amygdalas, one in the left and one in the right temporal lobe, but as with other paired structures in the brain, they are often referred to in the singular.) The amygdala plays an important role in sexual arousal and in other emotional states, such as fear. The fact that this patient experienced both emotions at the same time suggests that a malfunction of the amygdala played a role in his seizures. In fact, it's possible to trigger sexual arousal, and even orgasm, by direct electrical stimulation of the amygdala.[2]

A BRAIN NETWORK FOR
SEXUAL AROUSAL

No single region of the brain is its "sex center" or "fear center." Any given region performs a particular transformation of the signals it receives from other regions—a transformation that, viewed in isolation, is meaningless. The amygdala is part of a network of structures whose collective activity leads to sexual arousal, as I'll discuss. Another cautionary statement: most large structures within the brain comprise multiple subdivisions with differing connections and function. I will gloss over many of these complexities, which can be explored further in online reviews.

Sexual arousal has two components. The first consists of psychological states such as sexual excitement, the urge to engage in sexual behavior, and the pleasure associated with orgasm—I'll call this "mental arousal." The other consists of physiological phenomena such as lubrication of the vagina and erection of the penis. It includes other possible physiological changes, such as increases in heart and respiration rates, but it is the changes in the genitals that are usually measured, so I'll call this "genital arousal." These two kinds of arousal might be caused by the same or different mechanisms, or one kind might trigger the other. In the case history cited earlier, mental arousal occurred without genital arousal, suggesting that mental arousal is not simply a response to physiological changes in the genitals.

Figuring out the brain network responsible for sexual arousal is very much a work in progress. Still, we do know that arousal depends on brain processes that involve consciousness, as well as other processes that are entirely unconscious.

Sexual arousal can be triggered by stimuli arriving at the brain via any sensory modality—sight, sound, touch, taste, or smell—or by internal processes such as memory or fantasy. But most studies of how the brain produces sexual arousal have focused on one sense: sight. The brain's vision centrale is at the back of the cerebral cortex—the OCCIPITAL LOBE and nearby parts of the temporal lobe—so that's where our account of sexual arousal begins.

One informative study was conducted by a group led by Claus-Christian Carbon of the University of Bamberg, Germany, using what are called "event-related potentials" (ERPs).[3] These are electrical signals originating in the brain that occur in response to a triggering external event and can be recorded with an array of electrodes placed on the scalp. In this case, the event was a photograph of a face flashed onto a screen. The researchers asked the volunteers to press a button to indicate whether the face was male or female or (in separate trials) whether it was attractive or unattractive. For both tasks, it took the participants the best part of a second to press a button. By examining the recordings of brain activity, however, the researchers were able to show that the occipital lobe completed the task of determining the face's sex by less than a quarter of a second after the image was flashed. The analysis of attractiveness, however, didn't even begin until fifty milliseconds later. In other words, the brain already knows—or thinks it knows—a face's sex while it is analyzing attractiveness.

This finding is consistent with a considerable body of research showing that attractiveness is judged by sex-specific criteria. And it fits with common experience. Once in a while, most of us misjudge a stranger's sex. We may quickly realize our mistake, in which case that person's attractiveness is liable to change dramatically. I call that a "gender double-take."

Carbon's study was not about sexual arousal as such. Nevertheless, the likelihood that we will be sexually aroused by seeing a person's face depends greatly on that person's sex and attractiveness. Most of us are not aroused if the person is of the wrong sex ("wrong" as dictated by our sexual orientation, if we are attracted to only one sex), and we are not aroused by faces we judge to be unattractive. Thus events in the visual face and body areas are the initial steps in a process that, if other circumstances are favorable, may lead to sexual arousal.

The later steps have been studied using brain-imaging techniques, either functional magnetic resonance imaging (fMRI) or positron emission tomography (PET). Compared with event-related potentials, these imaging techniques are not nearly as good at analyzing the timing of brain activity, but they are better at identifying where activity occurs. (Although fMRI and PET are different technologies, the information they provide is similar, so I sometimes refer to brain scans or functional imaging without specifying the particular method.)

The first task, then, has been to map all the brain regions whose activity changes during sexual arousal, as these are candidates to be elements in a control network that generates and guides arousal. A great deal of brain activity is required to see anything, however, whether sexually arousing or not. Thus the imaging procedure is carried out in two conditions: with the participant viewing an erotic photograph or video and a comparable visual stimulus that is not erotic in nature (a sports scene, for example). To make visible the pattern of brain activity specifically related to sexual arousal, the images acquired in the nonerotic condition are digitally subtracted from the images acquired in the erotic condition, leaving—one hopes—just the activity that has to do with sexual arousal itself.

More than sixty studies of this kind have been published, and these have gathered data from a total of around two thousand volunteers of both sexes and varying sexual orientations. Most of the studies have assessed the volunteers' sexual arousal in some way. Experimenters simply may have asked them to report their mental arousal while viewing the images, or the researchers may have measured the volunteers' genital arousal by monitoring blood flow in their genitals. It's hard to imagine lying in the narrow confines of an MRI scanner while watching what is essentially porn and at the same time having an infrared camera trained on one's naked genitals—but that's the nature of sex research.

As you might expect, the findings of the various studies haven't been in complete agreement. The most meaningful conclusions are therefore drawn from meta-analyses, which combine the results from multiple studies following formal statistical procedures. Researchers at the Max Planck Institute for Biological Cybernetics in Tübingen, Germany, published one such meta-analysis in 2019.[4] The images shown in figure 3.1 are from that study.

The Tübingen study identified about a dozen active zones, but I am going to focus on seven of them. The first comprises several areas of the visual cortex that deal with the perception of faces and bodies. The others are two additional areas of the cerebral cortex—the anterior cingulate cortex and the inferior frontal gyrus—along with four subcortical structures: the amygdala, the HYPOTHALAMUS, the nucleus accumbens, and a region of the brainstem called the pons.

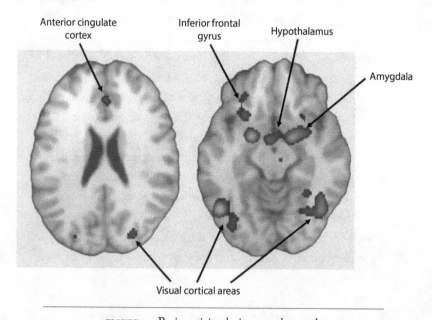

FIGURE 3.1. Brain activity during sexual arousal.

Source: From E. Mitricheva, et al., "Neural Substrates of Sexual Arousal Are Not Sex Dependent," *PNAS* 116 (2019): 15671–76.

These various structures constitute nodes in a network. The interconnections among the nodes can be assessed by tracing anatomical pathways—something that is usually done in laboratory animals—but also by physiological methods in humans that assess how much the activity in each node is dependent on the activity of other nodes. Such studies lead to a "wiring diagram," which is illustrated in simplified form in figure 3.2. This diagram is based in part on the work of a South Korean group led by Jin-Hun Sohn.[5]

The active areas of visual cortex include two neighboring zones, one dedicated to the perception of faces and another for bodies. They are located at the border between the temporal and occipital lobes in a region that is not directly visible in the midline view that is presented in the figure. These areas are probably where the rapid analysis of sex and attractiveness are performed, as reported in the ERP study discussed earlier. They become active during viewing of any faces or bodies, but the responses are much stronger during viewing of images that provoke sexual arousal.[6]

It's possible that the visual regions by themselves are capable of figuring out whether a stimulus should be arousing. It's likely, though, that higher centers, to be described later, send feedback signals to the visual cortex saying, in effect, "Hey, pay close attention to this" (if it's sexually arousing) or "Don't bother" (if it's not). These feedback connections are not shown in the diagram.

Outputs from the visual cortical areas go to the amygdala, among other places. Only part of the amygdala is involved in sexual arousal. Here, it is thought, information coming from the visual cortex and other sources is assigned an emotional value— a positive one, perhaps, if the visual signals relate to a potential sex partner who is judged to be attractive. Other parts of the amygdala deal with other emotions, such as anxiety or fear.

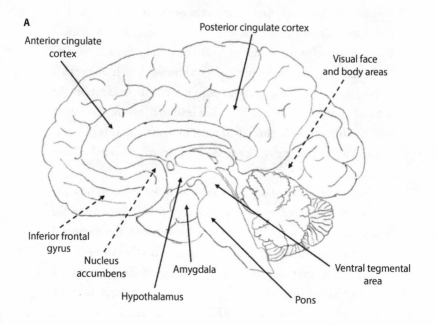

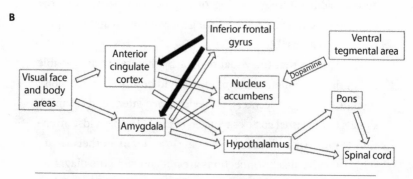

FIGURE 3.2. Brain pathways for sexual arousal.

Source: Figure by the author.

The amygdala also plays a role in memory. In particular, its activity ensures that we remember emotionally loaded events much better than events that are emotionally neutral. This seems to be true whether or not the event is a highly pleasurable sexual experience or a highly traumatic one, such as a sexual assault. The amygdala plays this role by virtue of its connections to the nearby hippocampus, the brain's major site for memory encoding.[7]

Outputs from the amygdala include a return pathway to the visual cortex, which is likely responsible for the enhancement of activity in the visual cortex when one views sexually arousing images, as described earlier.

Another output from the amygdala goes to cortical areas in the FRONTAL LOBE, including a region called the INFERIOR FRONTAL GYRUS, which is part of the PREFRONTAL CORTEX. This part of the brain is involved in higher cognitive functions, especially the initiation and control of behavior. You may desire to engage in some activity, such as having sex with an attractive stranger, but the prefrontal cortex decides whether such behavior would be appropriate in the circumstances. Most of the time, it's not. Thus the prefrontal cortex plays a censor-like role, which it exerts via inhibitory connections that run back to the amygdala. This inhibition damps down activity in the amygdala even when visual inputs, left to themselves, would be highly arousing.

Damage to the prefrontal cortex can impair this inhibitory function, leading to sexual behavior in inappropriate circumstances. The following account is from a case report of a seventy-five-year-old Italian man who suffered a fall that caused damage to the prefrontal cortical areas on the right side of the brain.

> He described himself as sexually inactive due to an erectile dysfunction but reported that when he was alone and encountered younger women, he felt a strong sexual desire and he was unable to stop himself from approaching them to make sexual advances. He also reported frequent intrusive thoughts about sexuality. He reported feeling stressed and ashamed by these impulsive behaviors because they were against his moral principles and habits,

which he described as very rigorous. He was a religious man and involved in several charity works of his parish.[8]

The anterior cingulate cortex is located just in front of the corpus callosum, the band of axons that interconnects the two hemispheres. Among a wide range of connections, it receives inhibitory input from the prefrontal cortex and excitatory input from the amygdala. (The latter pathway is not shown in figure 3.2.) The anterior cingulate cortex is involved in many cognitive processes, but no one has provided a satisfying explanation of its basic contribution to all of them. In the context of sexual arousal, it's been proposed that the anterior cingulate cortex generates a motivational element, directing behavior to a sexual goal if the circumstances are appropriate.

The hypothalamus, which is located on the undersurface of the brain, contributes to a variety of motivated behaviors that are considered basic to life, such as feeding and drinking and maintaining the correct body temperature. Small subregions within the hypothalamus serve each of those functions via outputs that ascend to the cerebral cortex, which trigger organized behavior (for example, going to a warm place when cold), as well as outputs that descend to the brainstem or spinal cord, which trigger more automatic activities such as shivering or sweating. The hypothalamus also controls the secretion of numerous hormones by the pituitary gland; this gland is tucked away in a cavity in the base of the skull immediately under the hypothalamus. Some of these hormones regulate the secretion of sex steroids such as testosterone and estrogen from the gonads (testes or ovaries). This gives the hypothalamus broad control over sexual development and sexual expression.

Several regions of the hypothalamus deal with sexuality. Two that I'll mention are the MEDIAL PREOPTIC AREA (or MPA), far forward in the hypothalamus, which is known to play a key role in male-typical sexual behavior; and the VENTROMEDIAL NUCLEUS (VMN), located farther back, which is required for female-typical sexual behavior as well as aggressive behavior by both sexes. (The word "nucleus," in the context of neuroanatomy, means a cluster of neurons that is consistently recognizable from brain to brain on the basis of its appearance or connections.)

The hypothalamus as a whole is so small, however, that it's difficult to distinguish between the subregions based on functional imaging methods such as fMRI and PET, especially when data are combined from many different studies. For that reason, the Tübingen study treated the hypothalamus as a single unit. Breaking down the roles of individual cell groups within the hypothalamus is best done in laboratory animals; I'll return to this topic later.

Another region that lights up during visual arousal is the NUCLEUS ACCUMBENS, which lies in front of the hypothalamus. The nucleus accumbens is the best-known part of the brain's reward system. Both laboratory animals and humans find electrical stimulation of the nucleus to be pleasurable—so much so that given the opportunity, they may stimulate themselves to the point of exhaustion while disregarding all other potentially rewarding activities.

Besides inputs relating to sexual and other emotionally loaded stimuli, the nucleus accumbens is a major target for axons from a region in the brainstem named the VENTRAL TEGMEN-TAL AREA. The endings of these axons release DOPAMINE within the nucleus accumbens. The activity in the nucleus accumbens may be responsible, at least in part, for the conscious pleasure that is commonly associated with sexual arousal and behavior.

Figure 3.2 doesn't show any outputs from the nucleus accumbens, but no structure in the brain is a dead end. In reality, the nucleus accumbens sends axons to at least twelve regions of the brain, many of them concerned in one way or another with the processing of emotions and instinctive behaviors.

This brings up an important question: what role does consciousness play in sexual arousal? Surprisingly, it turns out that much of the network can become active in response to an erotic image even when the person is completely unaware of what they are seeing. The evidence for this comes both from studies of persons with brain damage, as well as from studies in healthy persons who are made unaware of what they are seeing by psychological trickery.

Individuals whose primary visual cortex has been destroyed by a stroke have no conscious vision: whatever image is presented to them, they deny seeing it. Yet when such individuals are asked to guess what is being shown to them—for example, whether it is a smiling or an angry face—they can guess correctly at well above chance levels. And appropriate parts of the network become active, including the face and body areas and—if the image is arousing—the amygdala.[9]

Comparable observations have been made in people without brain damage by the use of subliminal stimulation. In one method, the image is shown very briefly and followed immediately by a "wipe-out" display of visual noise. Or the image can be presented to one eye at the same time as visual noise is presented to the other. In either case, the image is not consciously perceived; even so, the network for sexual arousal lights up in brain scans.[10]

TWO ROADS TO AROUSAL

The foregoing findings have led to the realization that there are two pathways by which visual stimuli activate the arousal network. One, which we may call the "high road," involves the primary visual cortex. This pathway can carry information about fine details of the image or video, and the activity of this pathway is strongly associated with conscious awareness.

The "low road," on the other hand, bypasses the primary visual cortex. Instead, it involves the activity of visual centers in the brainstem, whose output goes to cortical areas forward of the primary visual cortex, including the face and body areas, as well as to the amygdala and other parts of the network for sexual arousal. The low road conveys less detailed information, but the information that it does convey is transmitted extremely fast. This is a common phenomenon in the nervous system: detailed analyses are often preceded by quick-and-dirty ones, especially when it comes to behavioral situations that creatures have had to deal with since creatures existed, such as feeding, fighting, and sex.

The low road doesn't necessarily engage consciousness. Nevertheless, subliminal sexual stimuli can have all kinds of effects: facilitating genital arousal, inducing a more sex-positive mood, and biasing behavior toward sexual goals. In a study led by Omri Gillath of the University of Kansas, for example, heterosexual students were presented with subliminal images of naked persons of the other sex or with comparable images of dressed persons or abstract designs. Afterward, the students exposed to the sexual images were more likely to select a condom than a pen as a thank-you gift compared with students who viewed the other images—suggesting that they were more likely to be thinking about having sex.[11]

Although the visual information transmitted via the low road is not as detailed as that conveyed via the high road, it does carry enough information to allow for the distinction between images of males and females. A research group led by the psychologist Yi Jiang, formerly at the University of Minnesota, showed that volunteers paid more attention to subliminal images of naked persons belonging the sex to which they were attracted. For example, straight men paid more attention to images of naked women, whereas gay men paid more attention to images of naked men—even though the volunteers were not aware of seeing anything.[12]

In normal circumstances, both the high road and the low road are engaged in activating the sexual arousal network. But because of the speed with which the low road operates, it is likely that centers such as the amygdala are primed by unconscious low-road signals even before the detailed high-road information reaches them. This might lead to conflicts. What, for example, if the low-road message is interpreted unconsciously as "woman" but the later-arriving high-road message says "man"? Perhaps conflicts of this kind can trigger the gender double-takes that I mentioned earlier.

Of course, vision is not the only sense that can elicit sexual arousal. Probably any sense can do so, but one sense that has attracted particular interest is olfaction, the sense of smell. One belief that many people have is that olfaction is uniquely capable of drawing long-lost memories back into consciousness. The most-cited example is the "petite madeleine" of Marcel Proust's *In Search of Lost Time*—the little cake, dipped in herbal tea, whose taste and odor unspooled the narrator's childhood memories. Yet this belief about olfaction may not be correct. It's not so much that olfaction is better than other senses at triggering memories; rather, it's worse than other senses at everything else. In fact, in spite of the madeleine story, Proust represented all senses as being equally capable of triggering long-lost memories. (Later in the novel, he recounted episodes in which memories were triggered by the sound of a spoon knocked against a plate, the view of certain trees, the brush of a starched napkin across the lips, or the feeling of standing on uneven paving stones.)

Studies in laboratory animals such as mice have shown that odors that act as sex pheromones, or chemosignals, activate specific parts of the principal olfactory receiving area in the brain, which is named the "olfactory bulb." From here, signals are transmitted to a subregion of the amygdala and, from there, to the two subregions of the hypothalamus that I just mentioned: the medial preoptic area and the ventromedial nucleus.

As I discussed in the previous chapter, the subject of human sex pheromones is a controversial one. I mentioned two substances—AND (androstadienone) and EST (estratetraenol)—that are candidates to be chemosignals released by men and women, respectively, but I wrote that the psychological effects of AND and EST are not very impressive in humans, at least according to studies that have been done so far.

Even so, these two substances do light up the hypothalamus in functional images of volunteers who smell them, according to a team led by Ivanka Savic of Sweden's Karolinska Institute. What's more, they described a sex difference in responses: the hypothalamus was activated more by EST in heterosexual men and more by AND in heterosexual women.[13] A Dutch group replicated Savic's finding with respect to AND, though only when using a high concentration of the substance—a concentration that may not correspond to what a person would normally encounter.[14]

THE HYPOTHALAMUS: WHAT MICE TELL US

This book is about human sexuality, but it's worth making a detour into animal research here, because studies of the hypothalamus in laboratory animals are offering new insights into how sexual behavior is generated and controlled. I'll focus on studies that illustrate the application of modern neuroscience techniques that are difficult or impossible to apply to humans.

A problem that has long bedeviled study of the hypothalamus is not just its small size or the even smaller size of the individual nuclei within it, but also the fact that neurons with different functions are often intermixed in seemingly random patterns. Furthermore, axons originating in each nucleus have to pass through other nuclei to get to their destinations. Thus if one observes a specific loss of function after destruction of a given nucleus, that function could be attributed to any of the cell classes present in that nucleus, or even to neurons in some other nucleus whose axons happen to pass through the nucleus that was destroyed.

Enter neurogenetics. This discipline takes advantage of the fact that different classes of neurons in the brain are different on account of their genes. Not that they possess different genes— all cells in an organism possess the same genome—but different sets of those genes are active, or expressed, in each cell type. This is true both when the brain is assembling itself before birth and during postnatal life when the brain is functioning. The differences in GENE EXPRESSION control which connections the neurons make, which activity patterns they are capable of generating, which hormones or neurotransmitters they are sensitive to, and which neurotransmitters they themselves use to influence other neurons.

Once the patterns of gene expression are known, it is possible to identify and locate many of these classes of neurons within the hypothalamus and reveal their sensitivity to hormones and their connections to other types of neurons. More than that, researchers have developed a cookbook of genetic procedures that allow them to silence just one class of neuron, or make it active when otherwise it would have been silent, or record its activity when the animal is engaged in a range of behaviors. Most of these experiments are done in mice—specifically, mice that have been genetically engineered to make these kinds of experiments possible. It takes a leap of faith, obviously, to apply the findings to humans. Still, the hypothalamus is an ancient region of the brain whose functions haven't changed greatly over the course of mammalian evolution. As an example of the kind of research that is being done in this area, I'll describe a set of experiments reported in 2020 by a research group at New York University Medical Center led by the behavioral neuroscientist Dayu Lin, with Takashi Yamaguchi as the first author.[15]

It's long been known that when a male mouse meets an unfamiliar mouse, it follows one of two simple rules: if it's a female, mate; if it's a male, attack. The mating response is triggered by neurons in the medial preoptic area, the attack response by neurons in a portion of the ventromedial nucleus. (Its official but indigestible name is the "ventrolateral division of the ventromedial nucleus of the hypothalamus," or VMHvl.) But how do these two groups of neurons acquire their distinct executive functions?

The NYU group showed that the activity of both groups of cells is driven by inputs from the POSTERIOR DIVISION OF THE AMYGDALA, or PA (see figure 3.3). Within the PA are two classes of cells; one sends axons to the MPA, and the other sends axons to the VMN. The two groups of PA cells are indistinguishable when viewed under the microscope with conventional staining methods, but they differ in their patterns of gene expression. Taking advantage of this information, the researchers showed that PA neurons that send axons to the MPA were active as a male mouse approached and mounted a female. Neurogenetic methods allowed the researchers to activate or silence those cells at will; when activated, the mouse would drop whatever

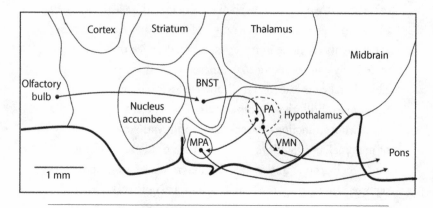

FIGURE 3.3. Hypothalamic pathways for sexual behavior.

Source: Figure by the author.

else it was doing and commence sexual behavior with a female. Conversely, when the cells were silenced, the mouse would not engage in sexual behavior even if a willing female sat right in front of its nose. In other words, activity in the pathway from the PA to the MPA is both necessary and sufficient for male sexual behavior with a female. Similarly, the neurons in the PA whose axons go to the VMN were active as a male mouse attacked an unfamiliar male, and this aggressive behavior could be switched on or off by activating or silencing those PA cells.

So there are two distinct channels from the PA to the hypothalamus in male mice that drive contrasting behaviors toward an unfamiliar mouse, depending on that mouse's sex. For this interaction to work, the two sets of PA neurons must "know" the sex of the other mouse. The precise source of this information isn't known with certainty. However, a study done at Stanford University with Daniel Bayless as the lead author identified a group of neurons named AB CELLS that, even in sexually inexperienced male mice, distinguish between males and females on the basis of their odor, becoming more active in the presence of females or when sniffing female urine.[16] AB cells are located in a structure named the BED NUCLEUS OF THE STRIA TERMINA-LIS, which receives inputs from the amygdala and also from the olfactory bulb.

When the Stanford group silenced just the AB cells, the mice no longer mated with females or attacked other males. Thus AB cells may provide the necessary excitatory input to the neurons in the posterior amygdala, which in turn drive activity in the MPA and VMN. This circuit operates only in male mice, however.

Female mice are not passive recipients of males' advances; they solicit sex from males by means of a display routine that includes hopping, darting, ear-wiggling, and ultrasonic vocalizing. If the male is sufficiently impressed by this performance, he will attempt to mount the female, grasping her flanks with his forepaws, whereupon she may raise her rump so as to expose her vulva and permit penetration—a behavior known as LORDOSIS.

The Stanford group used neurogenetic methods to activate the same neurons in the VMN of female mice as those that drive aggressive behavior in male mice. This caused the females to engage in sexual behavior—specifically, lordosis. It's surprising, perhaps, that the same group of cells drives different behaviors in male and female mice. Nevertheless, this finding is consistent with increasing evidence that the hypothalamus is wired differently in the two sexes. The medial preoptic area, for example, drives male-typical sexual behavior in males, but in females it is more concerned with maternal behaviors such as retrieving pups that wander away from the nest; the NYU group investigated the neuronal basis for this behavior in a separate study.[17] These sex differences are not absolute—in some circumstances males will retrieve pups and females will mount other mice or show male-typical aggression—but in the usual course of events, these are sex-differentiated behaviors that depend on wiring differences in the hypothalamus and elsewhere.

If the female is not motivated to engage in sex, she doesn't show lordosis, and there's little the male can do about it. The VMN cells that drive lordosis carry receptors for estrogen and require high levels of estrogen to respond to excitatory inputs. During the ovulatory portion of portions of a mouse's estrous cycle, when she is capable of becoming pregnant, her estrogen levels are high, so the VMN is responsive to excitatory signals. During the remainder of the cycle, her estrogen levels are low, and her VMN cells fail to trigger lordosis even if she is mounted by a male.

This is very different from what happens in humans, of course. Women (and the females of some closely related primates) can and do solicit and engage in sex even when they are not ovulating, as I discussed in chapter 1. So either this hypothalamic circuit is organized differently in women—making the VMN neurons less dependent on estrogen, for example—or other parts of a woman's brain can ignore or override her hypothalamus.

I've omitted mention of one cell group in the hypothalamus that's relevant to sexuality: the neurons that synthesize and release the hormones (or neuromodulators) oxytocin and vasopressin. Because these substances play an important role in pair bonding, I discuss them in the chapter on love (chapter 11).

THE PONS AND ORGASM

The final brain region in the arousal network that I'll mention is the PONS—the expanded portion of the brainstem to which the cerebellum attaches. A group led by Gert Holstege of the University of Groningen in the Netherlands has shown that a specific region within the pons is active in both men and women during orgasm and ejaculation but not during sexual excitement without orgasm or during faked orgasms.[18] They have named this region the PELVIC ORGAN-STIMULATING CENTER (POSC). In Holstege's study, the volunteers lay with their heads in a PET scanner while their partners stimulated their genitals, so it's likely that this region receives ascending inputs from the genitals. In addition, though, it's a target of connections descending from some of the forebrain regions that I discussed earlier, especially the medial preoptic area of the hypothalamus. The POSC is probably a key region in the triggering of orgasm. That doesn't mean that the POSC is where we subjectively *experience* orgasm: that experience likely reflects activity in an extended network of cortical and subcortical regions, including those discussed earlier.

As with most other brain structures, there are two POSCs, one on the left and one on the right side of the midline. But in Holstege's study, only the left POSC was active during orgasm or ejaculation. This was true in both the men and women who were studied. The right POSC, though silent during those sexual behaviors, was active during urination, as Holstege's group showed in a separate study.[19]

Urination is a nonsexual behavior (for most of us at least) and not an especially exciting one. Nevertheless, urination and male ejaculation have several features in common—both involve voiding substances through the urethra and the contraction of pelvic muscles. There are some similar functional asymmetries in the earlier parts of the arousal network, but not as clear-cut as this one.

CONCLUSIONS

Looking at the entire brain network for sexual arousal that I've described, we can draw some general conclusions. It's possible to trace a functional pathway that begins as a sensory process—namely, the visual analysis of faces and bodies—and then progresses through brain regions that attach emotional significance, pleasure, and motivation to the sensory stimuli, transforming them stepwise into behavioral triggers, as is seen in the hypothalamus and pons. Below the pons, centers in the spinal cord coordinate the actions of individual muscles to enable overt sexual behaviors.

Yet it's a mistake to think of this pathway as a simple forward progression, because it's really a broad network characterized by feedback and sideways connections. Thus even the first elements in the pathway—the visual centers that analyze faces and bodies—are subject to feedback influences from the amygdala and elsewhere that help shape neuronal activity. To really understand how sexual arousal happens, it will be necessary to study these interconnections in much greater detail than has been done so far and to replicate the interconnections in a digital model of the entire network.

It may never be possible to apply to humans the neurogenetic methods that in mice have clarified the hypothalamic circuitry for sexual behavior. But the resolution of MRI machines can be improved, most especially by increasing the strength of their magnetic fields. Already, the standard 3-tesla machines are being supplanted by new devices that operate at 7 tesla, and the most powerful MRI machine that a human has ever ventured inside, one that uses a 100-metric-ton magnet and 600 tons of iron shielding, operates at 10.5 tesla and has the potential to resolve a small fraction of a millimeter.[20] There's a lot that's waiting to be discovered.

I next move on to a specific aspect of human sexuality that has attracted a great deal of attention from scientists and laypeople alike: our sexual orientation. Too much attention, perhaps, especially during the many decades when the central question was, "What's wrong with gay people?" But that question having been definitively answered—"Nothing"—there still remains a mystery: how, and why, does it happen that people of the same sex can differ in such a basic characteristic as the sex or sexes that they find sexually attractive?

4

ORIENTATION

One afternoon in 2001, Harris Wofford, formerly the junior U.S. senator for the state of Pennsylvania, was swimming just offshore from a beach in Fort Lauderdale, Florida. On the shore stood an interior designer, Matthew Charlton, along with a friend. When Wofford came out of the water, the two men approached Wofford and began a conversation. Charlton and Wofford "clicked," and they later became companions and lovers.

This wasn't just another gay romance, however; it was unusual in two respects. First, the two men were fifty years apart in age: Charlton was twenty-five, Wofford seventy-five. When Wofford wrote about their relationship in the *New York Times* in 2016, shortly before he and Charlton married, many of the readers' comments focused on this age difference. A common thought was that a twenty-five-year-old could not be sexually or romantically attracted to a seventy-five-year-old, so Charlton must have been, as some readers uncharitably put it, a "gold-digger."

Relationships with large age differences are common in the gay community. I'm in one myself: my husband is nearly thirty years younger. So I wasn't inclined to doubt the mutual sincerity of the relationship between Charlton and Wofford. In fact, they were an item for nine years before same-sex marriage became legal in Washington D.C., where they lived; twelve years before the IRS recognized such unions; and fifteen years before they actually tied the knot. Furthermore, Wofford was in good health when the men met—as he later proved, by living to the age of ninety-two. In other words, the circumstances of their relationship were not such as to appeal to someone focused on a quick financial payoff.

I was surprised, however, by the other unusual aspect of Wofford and Charlton's relationship: Wofford wasn't gay. He had previously been in a forty-eight-year heterosexual marriage and fathered three children. The marriage was by all accounts a very loving one, and it was terminated only by his wife's death.

Of course, some men do come out as gay after a long heterosexual marriage—I've known several. When asked about their sexuality, such men will commonly say that they knew they were gay all along but didn't act on it for a variety of reasons. Some men engaged in same-sex relationships outside their marriages, with or without their wives' knowledge or acceptance. But this wasn't the case with Wofford. In his *New York Times* op-ed piece, he put it this way: "Too often, our society seeks to label people by pinning them on the wall—straight, gay, or in between. I don't categorize myself based on the gender of those I love. I had a half-century of marriage with a wonderful woman, and now am lucky for a second time to have found happiness."[1]

DO WOMEN HAVE A SEXUAL ORIENTATION?

What Wofford wrote about himself—"I don't categorize myself based on the gender of those I love"—was certainly unusual—for a man. But for women, statements of this kind are fairly common, and they're becoming commoner every year. Terms like "pansexual" (attracted to people without consideration of their sex or gender) and "omnisexual" (attracted to people of all sexes and genders but using different criteria for each) are crowding out the tired old categories of "straight," "gay," and even "bisexual," at least among socially conscious young women such as college students (as well as smaller numbers of male students). In fact, some sex researchers, such as the psychologist Michael Bailey of Northwestern University, have raised the question of whether women have a sexual orientation at all.

Before getting into those ideas, we need to agree on what we mean by SEXUAL ORIENTATION. If it simply means "the direction of a person's sexual attractions," it could encompass almost infinite possibilities: attraction to men, women, or nonhumans; to cis- or transgender; young or old; white, Black, or brown; fat or thin; hairy or smooth; cruel or kind; straight or gay; and to each in any degree or proportion. Everyone would have their own unique orientation, located at the intersection of all the dimensions that might have some bearing on their sexual attraction.

There's nothing unreasonable about using the term "sexual orientation" in that way, because all those dimensions of attraction are valid and worth understanding. But in both academic and common discourse, ever since the term was coined, it has been used to refer to just one dimension: the sex or sexes of the persons to whom one is sexually attracted. That's what the unqualified term "sexual orientation" means; other dimensions of attraction are sometimes specified by qualifying the same term, as in the phrase "age sexual orientation."

There's another issue that is still unresolved. When we talk about sexual orientation, do we mean how people name themselves—their stated identity? Or their stated sexual attractions?

Or the specific quality of their attractions—physical or romantic? Or their sexual arousal—arousal that's mental or physiological, where the latter most commonly means changes measurable in the genitals? Or their actual behaviors or relationships—at the present time or over their lifetime?

Among these options, genital arousal is the one that is most amenable to objective measurement in the laboratory. Erection of the penis can be measured with a strain-sensitive cuff; engorgement of the vulva and vagina can be measured by monitoring changes in temperature, the translucency of the vaginal walls, or the ratio of oxygenated to deoxygenated blood. Vaginal lubrication can also be measured using paper strips that wick fluid out of the vagina.

Bailey and his colleagues have made extensive use of such techniques to assess the sexual orientations of both women and men. Beyond the advantage of measurability, Bailey believes that sexual arousal is in principle closer to what we mean, or should mean, by sexual orientation, because it indicates that the brain or body is preparing to engage in sexual behavior. Attraction, on the other hand, could mean an intense friendship or romance that has no sexual component. And measuring genital arousal may circumvent a person's inability or unwillingness to put their sexual feelings into words or their use of terms, like "bisexual," whose intended meaning may not be completely clear to the researcher.

Even so, using arousal as a measure of orientation itself raises difficulties. Genital arousal can occur without attraction: a gay man may fully engage in sex with a woman, for example, without experiencing any sexual attraction to that partner. A woman may become genitally aroused while being sexually assaulted—not because she experiences any attraction to the rapist, but perhaps as a self-protective response: her brain may know, at some level, that she will be less likely to suffer injury if her genitals are prepared for sex.[2] Conversely, attraction can occur without genital arousal, most obviously in a man who suffers from erectile dysfunction. And when people are asked about sexual attraction, they understand it to mean a trait—their durable likelihood of experiencing attraction to men or women and not whatever attraction they may be feeling at the very moment of being asked. Genital measurements, on the other hand, assess a momentary state, one that's highly dependent on the details of the stimulus and other circumstances during the test.

In my view, the key element of sexual orientation is attraction, and the first and simplest way to assess that element is to ask a person to whom they are sexually attracted. Assessing genital arousal is less direct, although it does have the important merit of by-passing the psychological baggage that may complicate or impair a person's ability to give a valid response. Asking about behavior and relationships is also useful, but, of course, there are plenty of people whose relationships are at odds with their sexual attractions or do not represent the full range of attractions that the person experiences.

In surveys, most women say that they are sexually attracted only to men, just as most men say that they are attracted only to women. But more women than men express some degree of attraction to their own sex: in a U.S. national survey conducted between 2011 and 2015, 18 percent of women did so, compared with only 7 percent of men.[3] This difference was accounted for by greater numbers of women expressing attraction in some degree to *both* sexes—exclusive same-sex attraction was actually commoner among men than women. Recently, the numbers of women who identify as bisexual has been increasing, especially among women who are young, college-educated, nonwhite, or nonreligious.[4]

Another issue is whether women's sexual orientations change over time. Women who identify as heterosexual usually maintain that identity over the long term.[5] The same may not be true for women who claim a bisexual, lesbian, or other sexual-minority identity. When followed over a decade, such women commonly change their stated identity, though they usually stay within the umbrella category of "nonheterosexual."[6] Among the minority of women who do switch from a nonheterosexual to a heterosexual identity, most or all remain aware of some degree of sexual attraction to women, according to in-depth interviews conducted by the psychologist Lisa Diamond of the University of Utah.

Although Diamond has emphasized the "fluidity" of women's sexual orientation, based on her findings, this fluidity may relate to the terms women use more than to than their underlying attractions. In a 2022 UK study, changes in women's self-descriptors over time were not matched by equivalent changes in their genital arousal patterns.[7]

Laboratory studies of women's sexual arousal patterns do reveal significant differences from those of men. When men view erotic images of videos featuring men or women, the great majority report mental arousal to only one sex—heterosexual men to females, gay men to males—and their measured genital arousal matches these reports. Only some bisexual-identified men show significant penile erection to both sexes, and even in those men, genital responses are typically stronger to one sex than the other, according to Bailey's group.[8] Heterosexual women, in contrast, are nonspecific in their genital responses; they show genital arousal to erotic images or videos of both sexes. They may report that their mental arousal is only to males, but if they are led to believe that their statements are being monitored with a lie detector, they are more likely to report arousal to both sexes, in closer accordance with their genital responses.[9] In other words, in the normal situation (without the fictitious lie detector), their responses or their self-awareness may be modified by their sense of what is socially desirable. Perhaps men's erect penises act as built-in lie detectors, whereas women's arousal genitals are less visually prominent and thus can be more easily ignored.

It seems, then, that women's sexual arousal, both mental and genital, is less "target specific"—that is, less specific to the sex of the person in the image or video—than is the case for men. This is true only for heterosexual women, however; lesbians are more strongly aroused, both mentally and genitally, by female stimuli, and this is also true for some bisexual-identified women and even by some heterosexual-identified women who acknowledge a degree of attraction to females.[10]

In an attempt to assess people's sexual attractions more objectively than simply by asking them, several researchers have used a selective looking task, which depends on the fact that people tend to look longer at individuals they find attractive. For example, the psychologist Richard Lippa of California State University, Fullerton, had self-identified heterosexual, bisexual, and gay/lesbian men and women look through a series of photographs of "swimsuit models" of varying degrees of attractiveness. (The models' attractiveness had been assessed previously by a panel of student judges.)[11] The photographs were not explicitly erotic; they were the kinds of photos that one frequently sees in nonpornographic magazines. Lippa tasked the participants with placing the female models in rank order of attractiveness and also the male models. Without the participants' knowledge, Lippa recorded the length of time they spent looking at each photo.

The results were clear-cut, especially for men: heterosexual men rated female models more attractive than males; in fact they basically gave all males a zero for attractiveness. They also spent more time looking at the females, especially the most attractive females. Similarly, gay men rated the male models more attractive than the females, gave the females consistent zeroes, and looked longer at the males, especially the most attractive ones. Bisexual men rated male and female models about equally attractive and spent about the same time looking at males and females. This was somewhat contrary to what Bailey's group reported, in that those researchers had to narrow down the group of bisexual-identified men to find a subset who showed genital responses to both sexes, and even then, the responses were not usually balanced for the two sexes. Lippa, on the other hand, got a bisexual looking pattern for nearly all the bisexual-identified men who volunteered for the study. Men's sexual orientation, according to Lippa's measurements of sexual attraction, is just what men say it is, at least within the nonjudgmental environment of a psychology lab.

Lippa obtained somewhat similar results for women: heterosexual women rated male models more attractive than females, and they looked at the males longer. Lesbians showed the converse pattern, whereas bisexual women rated both males and females about equally and looked for comparable lengths of time at the two sexes. In contrast with the male results, however, the heterosexual women did not give zeroes to their nonpreferred sex—females; rather, they rated females only modestly less attractive than males, and they were strongly influenced in their ratings by the women's actual attractiveness. And although the heterosexual women looked at the females for less time than they did looking at the males, they spent much more time looking at the females than the heterosexual men spent looking at their nonpreferred sex—males. So, contrary to Bailey's earlier findings, Lippa's results indicate that both heterosexual and lesbian women do, indeed, have sexual orientations, but they are not nearly as clear-cut as the orientations of straight and gay men.

Where does that leave omnisexual and pansexual women and men, as well as those, like Harris Wofford, who simply reject categories and think only in terms of specific individuals who attract them? If the sexual orientations of such individuals were assessed with the kinds of tests that I've been describing, they would probably yield results similar to those for bisexual people. But omni- and pan-identified people would presumably also show attraction to a variety of trans and nonbinary persons, although I don't know of studies that have examined this. Bisexual people might or might not be attracted to trans and nonbinary persons—it's just not specified in the definition and hasn't been explored scientifically. And whether pansexual-identified individuals are truly distinct from omnisexual people—in not judging people's attractiveness by sex-specific criteria—also remains to be investigated in the laboratory. It's possible that people who claim some of the less common sexual identifications, of which there are many, are motivated by the desire to be different or by social contagion more than by any actual differences in their experience of sexual attraction. There's plenty of work still to be done to understand the diversity of sexual orientation, especially those of women.

One other point worth making: sexual orientation is not just about explaining homosexuality or any form of nonheterosexuality. It is also about how it happens that men, on the whole, are attracted to women and women to men, two sexual orientations that go by the same name—heterosexuality—but are actually radically different.

GENES

Andrea Ganna, an Italian geneticist, leads a research group at the Finnish Institute for Molecular Medicine in Helsinki and also has a position at the Broad Institute in Boston, Massachusetts.

In 2019, he was the lead author of a paper in *Science* titled "Large-Scale GWAS Reveals Insights Into the Genetic Architecture of Same-Sex Sexual Behavior."[12] This paper provided the most recent contribution to a long-running debate about homosexuality: is it genetic? Ganna's answer was "sort of," which some media accounts took as "yes" and others as "no."

Most gay people think of themselves as "born that way"— I know I do. No one remembers being born, of course, let alone what their sexual orientation was at the time, so this belief is of little real value in understanding the origin of a person's sexuality. Still, it does run counter to the notion, espoused by some antigay clerics and politicians, that homosexuality is nothing more than a choice people make. Choosing to be gay would be a life-altering decision that you'd not easily forget, even if you were blind drunk at the time you made it. So the fact that few, if any, gay people recall making such a decision strongly suggests that it didn't happen. Of course, there are many conceivable developmental pathways that don't involve either of these two extreme possibilities.

Long before Ganna's study, various lines of evidence indicated that genes are part but not all of the cause of a person's sexual orientation. Like most genetic traits, homosexuality runs in families—as does heterosexuality, for that matter—but there are also plenty of gay people who have no gay relatives, or none that they know of.

If someone who's gay has a same-sex twin, the chances that their twin is also gay depend on whether the twins are identical or fraternal. Identical twins, who have the same genomes, are far more likely to both be gay than fraternal twins, who are genetically no more similar than are regular siblings. This is consistent with a substantial genetic influence on sexual orientation, yet there are also many identical twin pairs in which one twin is gay and one straight. Such discordant pairs should not crop up at all if genes were the sole determinant of a person's sexuality.

Starting in the early 1990s, several groups of molecular geneticists began to search for the actual genes that might influence sexual orientation—"gay genes," as they were soon called. Several positive discoveries were reported, not for specific genes but for approximate locations in the genome where such genes were to be found. These included locations on the X chromosome and several other chromosomes. The results of the various studies were not in agreement, however; in retrospect, it's clear that they enrolled too few participants to yield definitive findings.

Ganna's group took advantage of two very large databases. One was derived from the UK Biobank survey, which has obtained personal and genomic information on more than four hundred thousand participants, both women and men. The personal information in the database was obtained by means of a questionnaire that asked, among many other questions, "Have you ever had sex with a same-sex partner?" About 4 percent of the male participants and 3 percent of the female participants answered "yes," and these individuals were asked further questions about how frequently they had had same-sex contact and with how many partners.

The genomic information was based on more than eight hundred thousand "markers," specific locations strung out along the entire genome where the DNA sequence may differ from one individual to another. The different DNA sequences at each of these locations are known as ALLELES. The other database was created by 23andMe, a company that offers health-related genetic information to the general public based on the same markers. About seventy thousand of their customers, mostly Americans, agreed to provide information on personal topics, including their history of same- and opposite-sex contacts and their attraction to males and females.

Many of the Biobank participants were related to each other to a greater or lesser degree, and Ganna took advantage of this information to derive a measure of the HERITABILITY of homosexual behavior: the percentage of the total causation of such behavior that could be attributed to genetic variability in the population. The answer was 32 percent, which was in the same ballpark as has been reported for homosexual attraction in twin studies. This result confirmed that genes have a substantial influence on sexual orientation, at least as far as behavior is concerned, but they are not the sole causal factor.

Ganna's group went on to examine the eight hundred thousand markers to see if there were any at which the alleles possessed by people who had had same-sex experience differed in a consistent way from those possessed by the other participants. The markers used in the study were locations where one specific BASE (nucleotide, or "letter") in the DNA code can be any of two or more alternatives: these markers are called SINGLE NUCLEOTIDE POLYMORPHISMS, or SNPs (pronounced "snips"). This is known as a genome-wide association study, or GWAS.

As expected, Ganna and his colleagues found a great deal of random variation, and that's why it took examining hundreds of thousands of people to see significant patterns. But they did, indeed, get some hits. At five locations in the genome—on chromosomes 4, 7, 11, 12, and 15—there were differences in the alleles of the two groups that met a strict criterion for statistical significance. (Humans have a total of twenty-three pairs of chromosomes.)

Of these five hits, two (on chromosomes 7 and 12) were linked to same-sex experience in both men and women, one was linked only in women (chromosome 4), and two were linked only in men (chromosomes 11 and 15). The researchers confirmed three of these hits, including both the male-specific hits and one of the both-sexes hits, in three smaller, independent data sets.

At none of the genomic locations identified by Ganna's group was the difference between the two groups consistent enough to account for more than a small part of the total heritability of the behavior. They certainly couldn't be made the basis for a meaningful test of a person's likelihood of engaging in homosexual contacts, although one company did try to market such a test. There must therefore be many more places on the genome where genes exist that have even smaller effects. In other words, there's no single gene that promotes same-sex behavior; rather, there's a large assembly of genes, each of which makes a small or tiny contribution. These genes may interact with each other in complex networks.

The five genomic locations identified by Ganna's group were different from the locations that had been reported in earlier studies. Those previous identifications may have been spurious—the results of random noise in the data—but they could also have been genuine hits that cropped up in those studies because of the particular populations that were studied, or because the participants were asked about different aspects of their sexual orientation.

It surprised me that Ganna found any markers linked to same-sex behavior in both sexes. After all, homosexuality in men and women is different: attraction to males in one case and to females in the other. It's really only because society judged same-sex behavior to be abnormal in both sexes that the two phenomena were given the same name. Perhaps the genes at the locations on chromosomes 7 and 12 contribute not to same-sex attraction as such but to some personality trait that makes same-sex behavior more desirable in either sex, such as novelty seeking, openness to experience, or rebelliousness.

Asking "Have you ever had sex with a same-sex partner?" hardly seems like an adequate way to probe a person's underlying sexual orientation, and Ganna specifically restricted his conclusions to same-sex behavior rather than attraction. Isn't it possible, after all, that many heterosexual people have had at least one same-sex contact, if only to see what it's like? Yet, the percentage of Biobank participants who ticked that box was no higher than the percentage of self-identified gay, lesbian, and bisexual people in the general population, which was 5.2 percent in a 2021 U.S. Gallup survey.[13] If most gay or bisexual people have experienced at least one same-sex contact, that doesn't leave much room for heterosexual women or men to join the party. I therefore believe that Ganna's findings are broadly relevant to sexual orientation and are not heavily contaminated by one-off sexual escapades.

What's more, when Ganna looked more closely at the Biobank and 23andMe data sets, it was clear that in males, at least, a same-sex contact is not usually a one-off event. If a man has done it at least once, then most of his sexual contacts will likely have been with other men. In contrast, a woman who's had at least one same-sex contact is likely to have had more opposite-sex than same-sex contacts. This goes along with the finding, mentioned earlier, that the majority of nonheterosexual men are gay, whereas the majority of nonheterosexual women are bisexual. The 23andMe data also indicate a strong correlation between ever having had same-sex contact with same-sex attraction and identification as gay, lesbian, or bisexual.

The fact that homosexual behavior, attraction, and identification are all part of a correlated package doesn't necessarily mean that genes are responsible for the packaging. That's because, as emphasized earlier, genes are less than half the total causation of homosexuality. The correlations among behavior, attraction, and identity might derive from shared causation of the nongenetic components. And, in fact, Ganna's group did not find strong evidence that the same set of genes both increases the likelihood of ever having had same-sex contact and promotes same-sex attraction or identification. Clearly, there's room for more work on the complex interactions between different aspects of a person's sexuality.

Identifying genomic markers linked to same-sex behavior doesn't mean that specific genes have been identified. A linked marker indicates only a general zone of interest—zones large enough to accommodate several candidate genes as well as DNA between the genes that also may play some role. Even so, Ganna's group did point out some interesting candidates. One of the markers associated with same-sex behavior in males—the one on chromosome 15—is near genes involved in the regulation of sex hormones and male-typical sexual development, and the other male-associated marker—on chromosome 11—is near a group of genes that code for olfactory receptors.

The Biobank consortium is engaged in sequencing the entire genomes of all four hundred thousand participants. When this herculean task is finished, it should be possible to identify the precise genes that were responsible for the five hits reported in Ganna's study, and perhaps other genes that the group was not able to identify. This in turn would enable researchers to investigate how specific variations in the DNA sequences make same-sex behavior more or less likely. Thus although the notion of a single "gay gene" is rapidly fading, there are likely to be important advances in our understanding of how genes contribute to sexual orientation at the behavioral level, and probably also at the levels of attraction and identity.

WHY DON'T GAY GENES DIE OUT?

The existence of genes that predispose their owners to same-sex attraction and behavior raises the question, "Why are they not eliminated over the generations?" After all, it's well documented that both gay men and lesbians have fewer children, on average, than their heterosexual peers.[14]

Many potential explanations for this paradox have been put forward, but two lead the pack in terms of actual support in the data. One, the so-called kin selection hypothesis, proposes that gay people who are not themselves parents nevertheless support the reproductive success of their close relatives, especially their siblings and their siblings' children. Because siblings have about half their genes in common, this seemingly altruistic behavior would actually help gay people perpetuate some of their own genes, albeit less efficiently than if they had been parents themselves. The Canadian psychologists Paul Vasey, Doug Vander-Laan, and their colleagues have reported finding such an effect among the *fa'afafine* of Samoa. These are a traditional "third gender" class of biological males who are psychologically feminine and sexually attracted to conventional men.[15] The researchers have documented that *fa'afafine* do indeed provide more support to their siblings' families than is provided by conventional men. Whether this support is enough to compensate the *fa'afafine* genetically for their own lack of children, and whether gay people around the world generally provide support of this kind, is not yet known.

The other idea behind the aforementioned paradox is the so-called fertile female hypothesis. This hypothesis is also based on the sharing of genes between relatives; the female relatives of gay men are predicted to be more fertile (produce more children) than women who don't have gay male relatives. Those same genes that make gay men attracted to males, when they are present in females, are thought to increase the women's reproductive success—by making them "hyperheterosexual," perhaps, or by making them more attractive to men. Again, then, the reduced reproductive success of the gay men themselves is thought to be compensated by the increased success of their relatives. Family studies in an Italian population have found a substantial effect of this kind,[16] but there have also been negative findings. This mechanism could also work in the reverse direction—helping preserve genes that promote homosexuality in women by increasing the reproductive success of their male relatives—but there are no data on this as far as I'm aware.

What's clear is that there are at least two and possibly several plausible mechanisms whereby genes promoting homosexuality could be maintained in the population despite homosexuality's reproductive cost. This topic is discussed in more detail in my book *Gay, Straight, and the Reason Why.*[17]

THE BRAIN

To a first approximation, the brains of men and women, as well as those of gay and straight people, are identical. With that important fact out of the way, let's focus on the subtle or not-so-subtle differences, some of which may be relevant to the development of sexual orientation.

An important pioneer in this field was the neuroscientist Roger Gorski, an emeritus professor at UCLA who died in 2021. Back in 1978, he and his colleagues reported finding a cell group in the rat's hypothalamus that was about five times larger in males than females.[18] The cell group was located in the medial preoptic area (MPA), the region that helps generate male-typical sex behavior, as I explained in the previous chapter. The cell group is now known as the SEXUALLY DIMORPHIC NUCLEUS OF THE PREOPTIC AREA, or SDN-POA. (The preoptic area is the general region at the front of the hypothalamus that includes the MPA.)

A fivefold size difference—or sexual dimorphism—is well into the not-so-subtle category; in fact, it can be seen in stained slices of the rat's hypothalamus even without the aid of a microscope, making me wonder why it wasn't spotted decades earlier. According to Gorski, the sex difference was first noticed by a postdoc, Larry Christensen, but for some reason, Christensen was not listed among the authors of the resulting paper.

Gorski, as well as other researchers, went on to investigate how this sex difference comes about. In brief, the cells that give rise to SDN-POA are generated during a rat's fetal life. A day or two before birth, they develop receptors for testosterone (androgen receptors) and become dependent on testosterone for their survival and further growth. The testes of male fetuses and newborns secrete large amounts of testosterone—plenty enough to keep the SDN-POA in good shape—but females produce much less of the hormone, and as a result, many of the neurons in a female's SDN-POA die. Artificially raising the levels of testosterone in a newborn female rat causes more cells to survive that usual, and her SND-POA ends up more like that of males. Conversely, depriving a newborn male of testosterone—by castration or drugs—causes many SDN-POA cells to die, and the nucleus ends up more like that of females. Castrating an adult male rat, on the other hand, does not change the size of SDN-POA, indicating that the cell group loses its dependence on testosterone at some point during postnatal male development.

At the same time, these early hormonal manipulations profoundly affect the sex behavior shown by rats in adulthood, leading to masculinized behaviors in females and feminized behaviors in males. This relationship between sex hormone levels during development and sexual behavior in adulthood was known well before Gorski's 1978 study. Pioneering behavioral experiments were carried out in the late 1950s by the neuroendocrinologist William Young and his colleagues at the University of Kansas. They used guinea pigs; similar studies were later carried out in primates.

In 1989, Gorski's group reported on their search for a SEXUALLY DIMORPHIC cell group in the human hypothalamus that might correspond to the rat's SDN-POA. The lead author of that study was Gorski's graduate student Laura Allen. They identified four cell groups in the preoptic area of the human hypothalamus that were candidates to be the human equivalents of the rat's SDN-POA. They called these cell groups the four "interstitial nuclei of the anterior hypothalamus," or INAH-1, 2, 3, and 4. After measuring the volumes of these cell groups in a large number of male and female brains, they found that INAH-3 was indeed sexually dimorphic, being about three times larger in men than in women, on average. They did not name this cell group SDN-POA, however, because a Dutch group had earlier given the name—erroneously, as it turned out—to another cell group, INAH-1. Thus INAH-3 retains its name but very likely does correspond to the rat's SDN-POA.

The sex difference in INAH-3 has been confirmed by several other laboratories, and it is about the only anatomical sex difference in the brain that is widely accepted, even by skeptics. Thus Rebecca Jordan-Young of Barnard College, in a 2010 book subtitled *The Flaws in the Science of Sex Differences*, acknowledged that the sex difference in INAH-3 "does appear to be real." But she immediately downplayed that concession by describing INAH-3 as a "little nucleus of unknown function" that might be related to "something as mundane and 'nonpsychological' as menstrua-tion."[19] That notion is inconsistent with a wealth of data, includ-ing the mouse studies described in the previous chapter, showing that in males, this part of the brain is concerned with directing sexual behavior toward females.

Similarly, a 2021, review titled "Dump the 'Dimorphism,'" by a group at Rosalind Franklin University School of Medicine and Science, acknowledged the sex difference in INAH-3 but dis-missed it as "tiny exception" that "[hasn't] been assigned a clear behavioral function."[20] Scientists know, however, that a tiny exception sometimes provides the key to new knowledge.

Despite the not-so-subtle size difference between the IHAH-3s of men and women, there is some overlap between the sexes, as if a person's sex is not the sole determinant of the cell group's size. In the discussion section of the UCLA paper, Laura Allen and her colleagues speculated that sexual orienta-tion might be another factor influencing the size of INAH-3 within each sex. Allen told me later that she had been keen to investigate this possibility but that Gorski, for whatever reason, was not on board with the idea.

That was when I jumped in. Although my area of specializa-tion was vision rather than sex, Allen's idea seemed too good to pass up, so I spent a year gathering and analyzing brains of deceased gay and straight men, as well as those of some women of unknown sexual orientation. INAH-3, as I reported in 1991, was about twice as large in the heterosexual men as in the gay men, on average.[21] Allen's hunch was correct.

A follow-up study by a group at Columbia University Col-lege of Physicians and Surgeons also found INAH-3 to be larger in straight than gay men, but the difference was smaller than what I reported, and its statistical significance was bor-derline.[22] That's the only study, of the more than two thousand papers and books that cited my paper, that included an attempt to replicate it. Clearly, a solid replication—or disreplication, if that's a word—is still needed thirty-one years after my paper was published.

More exciting to me was a 2004 study out of Oregon Health and Science University, led by Charles Roselli.[23] That group studied domestic sheep, a species in which about 5 percent of males (rams) have a strong and durable preference for sex with other rams rather than with ewes; in other words, they have a homosexual orientation. Sheep have a sexually dimorphic nucleus in the medial preoptic area; it is probably the equivalent of the human INAH-3 and the rat's SDN-POA. This cell group, Roselli reported, was half the size in the homosexual rams compared with the more numerous heterosexual rams. I was slightly shocked to learn that out there in nature was another species that offered such a close behavioral and neuroanatomical parallel to our own—at least as far as males are concerned. (Domestic sheep are not really "out there in nature," of course. It's possible that homosexual rams came about as a by-product of the process of domestication.)

The findings on INAH-3 are consistent with the so-called neurohormonal theory of sexual orientation, which proposes that sexual orientation is determined—or at least influenced— by interactions between sex hormones and the developing brain. This theory was put forward over a century ago by the German physician and gay rights pioneer Magnus Hirschfeld, and it has been bolstered by more recent research in animals, such as the experiments in rats mentioned earlier. The idea is that in fetuses that later become gay men, circulating levels of testosterone are unusually low, and this allows the brain to develop in a more female-typical direction. Alternatively, the levels of testosterone might be in the typical male range, but the brain may be less sensitive to testosterone than usual. The converse might be true for fetuses that become lesbian women: higher-than-usual prenatal testosterone levels, causing the brain to develop in a more male-typical direction.

Some measurable cognitive skills and personality traits in gay men are shifted, on average, toward values typical for women, and vice versa for lesbians.[24] In other words, scientific studies have shown that the standard stereotype about gay people—that gay men are feminine and lesbians are masculine—conceals a measurable kernel of truth. The neurohormonal theory offers a potential explanation for these gender shifts: atypical levels of prenatal testosterone could simultaneously affect the development of brain circuits for sexual attraction as well as those responsible for other gendered traits. I must emphasize the phrase "on average," though, because there's a great deal of variability in gay men's gendered traits, as there is among lesbians.

What, then, about transgender or transexual people? Is the size of their INAH-3 typical for their anatomical birth sex or for the sex with which they identify? The answer is the latter, at least for trans women, according to a study from the Netherlands Institute for Neuroscience. The INAH-3 in trans women was about half the size of the same cell group in cisgender men.[25] This finding is consistent with the neurohormonal theory.

The interpretation of the Dutch study is complicated, however, because most of the women had undergone hormonal treatments, which may have affected the size of INAH-3. In addition, while many trans women are sexually attracted to males (that is, they are homosexual with respect to their birth sex), some are attracted to females. So you can't say that transgender people are simply gay people taken to an extreme. Even though there are some people who occupy a kind of no-man's-land between homosexuality and transgenderism or transexuality, the great majority of gay people have a secure gender identity that is congruent with their birth sex.

Researchers using brain-scanning technologies have reported on a considerable number of differences in brain structure between men and women, as well as between gay and straight people of the same sex. Many of these reported differences have been small in magnitude, have been based on small numbers of individuals, or have not been replicated from one study to another. Such difficulties have provoked critiques such as the "dump the dimorphism" review mentioned earlier.

Big data studies have lent some respectability to the field, however. One such study, published in 2021, was led by Christoph Abé of Sweden's Karolinska Institute and Qazi Rahman of Kings College London.[26] This group analyzed data from more than eighteen thousand participants in the UK Biobank survey who, besides being questioned about their same-sex sexual experiences and having their DNA analyzed, as described in the previous section, also underwent MRI scans of their brains. Using multivariate analysis, a statistical procedure that is sensitive to the interaction of multiple variables, the Swedish-UK group found a significant pattern of differences between men and women. In those men and women who reported having had sex with a same-sex partner, however, these anatomical sex differences were less marked. In other words, the patterns of brain structure in these individuals were shifted partway in a sex-atypical direction, and this was a statistically robust finding.

To be fair, the dimorphism that the Rosalind Franklin University group say should be dumped is different from what is commonly meant by the term. In its original sense, sexual dimorphism referred to the radical differences between the sexes that one sees, for example, between peacocks and peahens. These days, though, it is generally used to mean any measurable difference between the sexes, even if there is overlap between the two. It may be that the authors of the "dump the dimorphism" study appealed to the older usage simply because it was so easy to shoot down.

There have also been functional MRI (fMRI) studies comparing the brain responses to sexually arousing visual stimuli in gay and straight people. Not surprisingly, the activity patterns are generally similar, as long as the stimuli are appropriate to the participants' sexual orientation. That is true, for example, for the hypothalamus, at least in men: whatever videos a straight or gay man finds sexually arousing causes the hypothalamus to light up, according to an fMRI study led by Thomas Paul of University Hospital, Essen, Germany.[27] The scanning technology used by that group was not precise enough to allow for the identification of specific active zones within the hypothalamus.

More interesting, perhaps, is a set of studies that employed olfactory rather than visual stimuli. In chapter 2, I mentioned two substances that may act as sex pheromones, or at least influence a person's disposition toward romantic interactions: androstadienone (AND), secreted by men, and estratetraenol (EST), secreted by women (and excreted in their urine). Ivanka Savic and her colleagues at the Karolinska Institute reported that exposure to AND activated the front portion of the hypothalamus in gay men and heterosexual women but not in heterosexual men or lesbians.[28] In contrast, exposure to EST activated this same region in heterosexual men but not in gay men, heterosexual women, or lesbians. The lack of activation in lesbians is surprising; evidently there is no simple reciprocal activation pattern between lesbian and straight women, as there is between gay and straight men.

Savic's studies have been criticized on various grounds, including the fact that the subjects were exposed to concentrations of

AND and EST that were higher than one would likely experience in regular life. In addition, as Savic acknowledged, it's not clear how the difference in activation patterns between gay and straight men (or between straight men and straight women) came about. It could have resulted from a difference in the process of wiring up the brain during early development—in which case, it may have been a factor helping to establish a person's sexual attractions. Alternatively, the different patterns could have been the consequence of many sexual contacts with same-sex or other-sex partners. This kind of chicken-or-egg problem is common to most studies that focus on adults, who are likely to be sexually experienced. Ideally, one would study young people before they become sexually active.

Of course, many other parts of the body differ to a greater or lesser degree between men and women, and these differences are brought about at least partly by differences in sex hormone levels during pre- and postnatal development. Thus if the neuro-hormonal theory of sexual orientation is correct, it's worth asking whether there are anatomical differences between gay and straight people in those other structures. Such differences have, in fact, been reported for several body parts. Here, I focus on two: fingers and faces.

There's a sex difference in the ratios of the lengths of the different fingers. In women, the lengths of the second and fourth digits are about the same, whereas in men, the second digit tends to be shorter than the fourth. Thus the so-called 2D:4D RATIO is typically lower in men than women, and evidence suggests that the difference results from the action of testosterone on the developing hand.[29] In 2000, a group led by Cynthia Jordan and Marc Breedlove, then at the University of California, Berkeley, reported that the 2D:4D ratio in lesbians is lower than in heterosexual women. In other words, it is shifted toward values more typical for men. This finding has been replicated in numerous studies.[30] Studies comparing 2D:4D ratios in gay and straight men have not come up with consistent findings (but see the following).

There are structural differences between the faces of men and women, such that people do a good but not perfect job of distinguishing male and female faces even in the absence of facial hair or cultural cues such as makeup or hairstyles. People can also distinguish between faces of gay and straight people, but with much lower accuracy—sometimes barely better than would be expected on a chance basis. In part, the structural differences are sex-atypical ones: for example, gay men, like women, have shorter noses than straight men, according to a 2014 study from the Czech Republic, yet this same study found some differences that did not fit the pattern.[31]

Investigations of this type are generally based on limited numbers of participants. The Czech study, for example, employed photographs of forty gay and forty straight men. But Yilun Wang and Michal Kosinski, of Stanford University's Business School, took a big data approach. In a study published in 2018, they trained a neural network on several tens of thousands of faces of both sexes, half of them gay and half straight.[32] (They obtained the photos from dating websites.) After training, the network was tested on pairs of previously unseen faces—pairs in which one of the faces was that of a gay person and the other of a straight person of the same sex. The network performed much better than human judges who were tested on the same faces, and its performance improved further if it was shown more than one photo of each face. In fact, when shown five different photos of each face, the network chose correctly 91 percent of the time when viewing male pairs and 83 percent of the time when viewing female pairs. (Random choices would be correct about 50 percent of the time.)

Wang and Kosinski showed that the network performed well even when relying only on the overall shape of the face—an anatomical characteristic that is largely independent of cultural factors. What's more, the network appeared to pick out the gay faces based on shifts in facial measurements toward those typical for the other sex. A similar conclusion was reached in a smaller study conducted by psychologists at Brock University in Canada, which found, for example, that gay men, like straight women, have shorter noses than straight men.[33] In other words, it appears that facial structure develops in a somewhat sex-atypical fashion in gay people—a conclusion that is consistent with the neurohormonal theory of sexual orientation.

Some commentators have warned that studies of this kind could lead to misuse, and that's certainly a concern. It's worth stressing, though, that Wang and Kosinski's network would be hopeless at picking, say, the gay individuals among passengers deplaning at an airport. That's because of the low percentage of the population that is gay: the correct picks would be swamped by false positives. It's only in the artificial one-of-a-pair situation that the network does a good job.

The most obvious anatomical differences between females and males are in the genitals; in fact, genital appearance is the usual basis on which a baby is assigned to the female or male sex. So in the simplest neurohormonal theory of sexual orientation, one might imagine that the penises of gay men may be smaller than those of straight men and that the clitorises of lesbians might be longer than those of straight women. Yet, no studies have reported such differences; in fact, a couple of studies have reported that the penises of gay men are *longer* than those of straight men.[34] I'm a little skeptical of those reports, but whether or not they are correct, they don't support the simplest neurohormonal theory. Most likely that's because genital structures differentiate earlier in fetal life than does the brain; the hormonal differences required by the neurohormonal theory of sexual orientation would therefore most likely exist later in fetal life.

There's another line of research that may be relevant to biological theories of sexual orientation. This has to do with birth order. According to a long series of studies by Canadian sex researchers—Ray Blanchard, Anthony Bogaert, and others—men who have an older brother are more likely to be gay than men who don't, and the more older brothers a man has, the more likely he is to be gay.[35] A cautionary note, however: in 2021, a group of psychologists at the University of Vienna claimed to find statistical errors in Blanchard's studies, errors that, when corrected, leave little evidence for a birth order effect.[36]

The birth order effect—assuming that it's real—could in principle result from the experience of growing up with one or more older brothers, but the Canadian group believes it has ruled out that possibility. The effect, according to a study by Bogaert, applies even to men who never lived with their biological brothers, whereas it does not apply to adoptees whose older brothers are not their biological relatives.[37]

The researchers have therefore proposed a biological explanation for the birth order effect. A woman who is carrying a male fetus may develop antimale antibodies, according to this hypothesis, and these antibodies persist in the mother's blood, affecting the sexual differentiation of the brains of subsequent male fetuses. This in turn raises the likelihood of same-sex attraction in adult life. In a partial test of this hypothesis, Blanchard, Bogaert, and several colleagues reported that mothers of gay sons are more likely than other mothers to possess antibodies against a male-specific molecule that may be involved in brain development.[38]

In conflict somewhat with this explanation, however, are the results of an Australian study published in 2022, with Christina Ablaza of the University of Queensland as first author. This group analyzed nine million marriages in the Netherlands, of which about sixty thousand were between same-sex couples—presumably gay, lesbian, or bisexual individuals. They found that having older brothers increased the likelihood that men would enter a same-sex marriage, consistent with the earlier results, but women with older brothers were also more likely to enter a same-sex marriage even though they don't possess male-specific antibodies. The effect of older brothers on sisters was not as strong as on brothers, but it was statistically robust. Blanchard has tweaked his theory to account for this result; it could also be, however, that different mechanisms are responsible for the older-brother effect in men and women, or the antimale antibody hypothesis simply may be wrong.

A direct test of the neurohormonal theory would involve measuring testosterone levels in fetuses and then following the resulting children through to puberty and beyond, when they could tell us about their sexual attractions. Such a study would be very difficult to carry out, but some "experiments of nature" provide information that's not far short of a direct test. One such experiment is the medical condition known as CONGENITAL ADRENAL HYPERPLASIA (CAH), in which the levels of testosterone and similar male-typical sex hormones (ANDROGENS) are unusually high during the later stages of fetal life. Women with CAH, who experienced these high androgen levels when they were fetuses, are much more likely to be lesbian or bisexual than women in the general population.[39] Even so, many women with CAH, even with the severest form of the condition, are heterosexual. It's as if prenatal hormones are a strong influence on sexual orientation, but some other, unknown influences must also exist for a woman to grow up lesbian.

Those other influences could be anything from genes to environmental factors such as parenting modes, or even sexual experiences in adolescence or adulthood. There is little direct evidence to identify any such environmental factors, however. In fact, studies of children whose biological birth sex was male but who were brought up as female—or the reverse—have generally found that birth sex trumps sex of rearing.

Another possibility is that prenatal EPIGENETIC EFFECTS play a role. These are biological processes that chemically modify genes, such that a gene with the same DNA sequence may function differently from one fetus to another. A UCLA group has presented preliminary evidence supporting a role for epigenetic effects in the development of sexual orientation.[40]

As I mentioned earlier, the prenatal hormonal theory could involve either atypical levels of the sex hormones themselves or something atypical in the brain's response to sex hormones or in the complex processes that lead from hormones to sexually differentiated neural systems. Marc Breedlove has argued that the former mechanism—atypical hormone levels—is responsible for homosexuality in women, while the latter mechanism—differences intrinsic to the brain—is responsible for homosexuality in men.[41] That may be true in many cases, but it can hardly be universally true. Recall that in Wang and Kosinski's big data study, gay men's facial structure was shifted in a female-typical direction. This can hardly be explained in terms of a process intrinsic to the brain; rather, it strongly implies that there was a difference between the average gay man and the average straight man in terms of the levels of sex hormones to which their faces were exposed during development.

This mention of averages is a good segue to an important point: gay men are not all alike, nor are all lesbians alike, and the differences among them may reflect different developmental processes. Breedlove made relevant observations during his study of finger-length ratios. After completing the initial study, he and several colleagues—Windy Brown was the first author— reported that the 2D:4D ratios of self-identified "butch" lesbians were smaller (i.e., shifted in a male-typical direction) than those of self-identified "femme" lesbians. In fact, the 2D:4D ratios of the femme lesbians were no different from those of heterosexual women.[42] The researchers concluded that only butch lesbians were exposed to atypical sex hormone levels during development. How the femme lesbians acquired their sexual orientation remained unexplained, but the researchers raised the possibility that social factors may have played a role. It is widely believed— though not clearly documented—that femme lesbians are more fluid in their sexual orientation than those who identify as butch, and that would be consistent with a less biologically controlling development. (There are plenty of lesbians who don't identify as either butch or femme.)

More recently, Breedlove's group (with Ashlyn Swift-Gallant as first author) reported analogous findings in gay men.[43] In this study, gay men who were "bottoms"—who preferred the receptive role in anal sex—had 2D:4D ratios that were greater (i.e., shifted in a female-typical direction) than "tops"—gay men who preferred the insertive role. Similarly, the bottoms reported a greater degree of gender nonconformity during their childhood compared with tops—a finding that has been reported in earlier studies by others. (Again, many gay men do not identify as top or bottom.)

CONCLUSIONS

The research discussed in this chapter supports the idea that sexual orientation can be best understood in terms of biological processes during development. More specifically, there appear to be differences between gay and straight people in the way sex hormones interacted with their brains during prenatal life. This is the neurohormonal theory of sexual orientation.

Of course, the theory is not an ultimate explanation of sexual orientation, because it raises questions like these: "Why are prenatal sex hormone levels different between different fetuses?" or "Why do some brains respond differently to sex hormones than others?" No doubt genes are part of the answer; that is certainly true in the case of nonheterosexual women with CAH, for example, because CAH is a genetic condition. Further research, such as the continuing analysis of the UK Biobank data, will likely provide much more information about how genetic differences between fetuses play into the neurohormonal processes discussed in this chapter. But it's clear at this point that for most people, there is no single gene that operates as a gay/straight switch. Rather, as with many other aspects of our mental lives, large numbers of genes, each with a small effect, interact with each other and with the brain in complex networks. These networks will not be easy to unravel.

Several lines of evidence, such as the findings on finger-length ratios, suggest that neither lesbians nor gay men are all of a kind in terms of their development; rather, they include a mix of individuals who experienced gender-atypical androgen levels during fetal life as well as other individuals who did not. Why, then, are the latter group not heterosexual? The answer is not yet clear. It could be that those individuals are gay because of differences in the way their brain responded to sex hormones or because of other biological processes that have nothing to do with sex hormones. It is also possible that social factors play a more important role than biological ones in their psychosexual development.

Given the known diversity among nonheterosexual women and men, the task of clearly identifying different "types" of non-heterosexuality—by multivariate analysis of the many factors that are now known to correlate with sexual orientation—and the linkage of such types to genes, hormones, and other biological markers will greatly improve our understanding of the topic.

Scientists like to keep things as simple as possible. Relying on a gay/straight or gay/bi/straight classification has been a highly productive strategy over the years. Still, it's also necessary to acknowledge the complexities that have been pointed out by critics of the biological approach. We haven't yet explained the Harris Woffords of this world.

Given the title of this book, some readers may be wondering if and when I am ever going to get to "sex" in the sense that it's commonly understood; namely having sex or "getting it on." The next chapter makes amends, perhaps, by tackling some of the intricate details of actual sexual encounters.

5

HAVING SEX

S ex should be fun, but scientists do their best to ruin it.
    I already described some of the things that research-
ers inflict on people while they are engaged in various kinds of sex: placing cuffs around their penises or optical
probes in their vaginas, for example, or putting them inside
MRI machines while arousing them sexually with pornographic
videos. But there's more.

One only mildly invasive study was carried out at Harvard
University by the psychologists Matthew Killingsworth and
Daniel Gilbert.[1] Killingsworth, who is now at the University of
Pennsylvania's Wharton School, is not a sex researcher as such;
rather, he studies human happiness. He and Gilbert devised a
project to assess people's instantaneous happiness—that is, their
degree of happiness at the moment of being asked. To that end,
they developed an iPhone app that called volunteers at random
times and asked them what they were doing and how happy they
were. People don't spend a large portion of their day having sex,
but occasionally the phone call did come at that inopportune
time. You might think that they would be unhappy—who likes
to be interrupted during lovemaking, after all? Yet, they reported
being extremely happy, far happier than during any other activity,
in fact. Maybe they reported on their mood just prior to the call.

People enjoy sex, then. And so they should, because the only purpose of human existence, from an evolutionary perspective, is to make more humans. It's no surprise that we are wired to find sex rewarding, as I discussed in chapter 1.

But there's more information in the Harvard study than that. Killingsworth and Gilbert, or their automated avatars, also asked volunteers whether they were thinking about anything other than the activity they were engaged in—whether their minds were wandering, in other words. That was often the case, and such mind-wandering was associated with a significant decrease in happiness. But lovemaking stood out for the infrequency of mind-wandering; most of the volunteers were really focused on what they were doing: they were in the moment.

Sex therapists believe that mind-wandering during sex, when it does occur, is one of the most frequent causes of sexual dysfunctions and relationship problems.[2] For clients experiencing such difficulties, they often recommend exercises known as "sensate focus." The idea is to pay attention to the sensations experienced during sex, to the exclusion of distracting thoughts such as, "What is my partner thinking about this?" or "I wonder how my retirement account did today?" Of course, it's probably a two-way street: distracting thoughts decrease enjoyment and thereby impair relationships; but, conversely, relationship difficulties promote distracting thoughts, such as "I'd rather be with _____."

ORGASM

So let's stay focused on sex. The hands-on study of this topic was famously pioneered by Masters and Johnson—the gynecologist William H. Masters of Washington University in St. Louis and

his assistant, later his wife (and even later his ex-wife), Virginia
E. Johnson. They are best known for their four-phase model of
human sexual response: excitement, plateau, orgasm, and resolution. I'm going to discuss just one of these phases: orgasm.

Orgasm has two elements: physiological processes that can
be measured and subjective sensations and emotions that require
description by the person experiencing them. Masters and Johnson are renowned for their physiological studies, for which they
recruited some extraordinarily long-suffering volunteers. How
many men, for example, would be willing to have the size, angle,
and position of their testicles repeatedly measured during their
sexual response cycle, including a few moments before orgasm?
How many women would tolerate being penetrated by an elec-
trically powered transparent penis, though which an investiga-
tor monitored the "sweating" of their vagina mucosa? Plenty, it
seems, because Masters and Johnson refer to their volunteers in
the hundreds, if not thousands—although they kept no records
and have been accused of exaggeration, even by Johnson herself.

Masters and Johnson maintained that there is only one kind
orgasm in women, whether or not it is triggered by stimulation of the vagina, clitoris, or some other part of the body. They
asserted that the clitoris is necessarily stimulated during vaginal
intercourse, because the hood of the clitoris rides over the clitoris
as it is tugged on by the inner labia. Thus, according to Masters
and Johnson, neither the size or anatomical position of the clitoris nor the coital position adopted by a heterosexual couple has
any significant bearing on whether the clitoris is stimulated during vaginal sex—it is always stimulated, and that's how a woman
reaches orgasm. In addition, the deepest internal portions of the
clitoris, which Masters and Johnson did not investigate, lie close
to the sides of the vagina, providing another means by which the
clitoris might be stimulated during vaginal penetration.

The fact remains, though, that many women don't experience orgasm during vaginal penetration unless they simultaneously stimulate the clitoris by hand, which can be awkward to do during missionary-position sex. In a 2014 study, a group led by the gynecologist Susan Oakley of the Good Samaritan Hospital in Cincinnati measured the size and position of the clitoris in women who did or did not experience orgasm during vaginal intercourse; Oakley used MRI scans and "blind" analysis rather than simple observation to gather data.[3] The group reported, in contradiction to Masters and Johnson, that the clitorises of the "anorgasmic" women were significantly smaller and positioned farther from the vagina than those of the women who could reach orgasm during vaginal intercourse. Similar findings were reported in a study of Turkish women published in 2021.[4]

This relationship between the position of the clitoris and the ability to experience orgasm during vaginal intercourse had actually been reported long before Oakley's study, according to historical research by Kim Wallen of Emory University and Elisabeth Lloyd of Indiana University.[5] The first study of this kind was conducted by a wealthy French psychoanalyst, Princess Marie Bonaparte, whose aristocratic credentials included being the great-grandniece of Napoleon, the wife of Prince George of Greece, the aunt of Britain's Prince Philip, and the mistress of the French prime minister Aristide Briand. Under a pseudonym, in 1924, Bonaparte published a paper in which she described measuring the vulvas of forty-three women. After establishing that women whose clitoris was farther from the vagina were less likely to experience orgasm during vaginal sex, Bonaparte went on to devise an operation to reduce this distance. She underwent the operation herself but did not derive any benefit from it.

In 1988, the New York–based sex therapist Edward Eichel, along with two associates, reported on a technique by which a man could directly stimulate a woman's clitoris with his penis while at the same time penetrating her vagina.[6] In Eichel's "coital alignment technique," the man lies directly on top of the woman without supporting himself with his arms and pulls himself forward so that his penis enters the vagina in a backward-pointing direction, sliding across the woman's mons and clitoral hood as it does so. Only shallow penetration of the vagina is possible in this position. The man and woman are supposed to engage in a reciprocal sliding motion during intercourse, so that his penis "plays" her mons and clitoris in the same way that a violinist's bow plays the strings of the violin. According to Eichel, many women who were unable to achieve orgasm during vaginal intercourse were able to do after adopting the coital alignment technique.

Subsequent studies by others have yielded mixed results: Some have verified Eichel's claims, whereas others have failed to do so, and a 2018 review was lukewarm about its value.[7] Another problem is that the average American man has packed on the pounds since the 1980s and is not just overweight but borderline obese, with an average weight of 197 pounds. A woman supporting such a weight without assistance from the man would likely be focused more on survival than reaching orgasm. (In fact, contemporary magazine accounts of the coital alignment technique advise the man to support some of his own weight on his elbows or hands.) Finally, while the coital alignment technique may improve some women's chances of reaching orgasm during vaginal sex, it may do the opposite for some men—that is, men who desire or need deep penetration of the vagina in order to climax.

All in all, it's clear that stimulation of the clitoris is vital for most women to reach orgasm, and an inability to reach orgasm during vaginal sex is often caused by a lack of adequate clitoral stimulation despite the ways in which stimulation may reach the clitoris during vaginal sex as described by Masters and Johnson.

This, of course, raises the question of why evolution placed the sensitive external portion of the clitoris where it did. According to a 2006 book by Elisabeth Lloyd, the clitoris has no adaptive function; it exist, she maintained, only because its male equivalent—the penis—needs to exist.[8] An accidental by-product, in other words, though a happy one. Since the time of Lloyd's book, plenty of counterarguments have been presented. For example, it might be that the position of the clitoris evolved to facilitate sex between women (think bonobos). Or, to invent a thoroughly implausible hypothesis, it's possible that evolution anticipated the invention of the personal vibrator, which makes it easy to stimulate the clitoris during vaginal sex, especially if the couple adopts positions such as the "cowgirl," which give either partner ready access to it.

In spite of the general emphasis on the clitoris as the trigger zone for female orgasm, some researchers kept their focus on the vagina—specifically, on the famous and highly controversial G-spot. This erogenous zone, if it exists, is located on the front wall of the vagina a few centimeters from its opening. According to a 1982 best seller by Alice Kahn Ladas, Beverly Whipple, and John Perry, stimulation of the G-spot is performed most easily by a come-hither motion of a finger, which can elicit an orgasm that is different in quality from a clitoral orgasm.[9] Specifically, a G-spot orgasm is described as "deeper" "throbbing," and some-times "more intense" than a clitoral orgasm.

There's nothing special about this region of the vagina that can be seen with the naked eye, under the microscope, or in MRI scans. (As with everything else about the G-spot, this statement is hotly contested.) Pressure on this region may stimulate under-lying structures, however, including the paraurethral glands, whose role in female ejaculation will be discussed later. Some inner parts of the clitoris are also possible targets of G-spot stimulation. Still, the biological basis of the G-spot, and even whether it exists at all, are unresolved questions. The G-spot has a much more secure reputation in women's and men's magazines than it does in the academic literature.

While on the topic of popular magazines, it's worth mentioning that almost every woman's magazine has at one time or another run a story about cervical orgasms. Many women find that stimulation of the cervix—by a penis, a finger, or a sex toy—is painful, but some do report experiencing pleasure and, on occasion, orgasms. These are said to be different in quality from G-spot or clitoral orgasms, and though accounts vary, the women all describe feelings that are even more "orgasmic" that regular orgasms: "It starts in the pelvis, spreads to the abdomen, and then engulfs the whole body" (*Glamour*); "an uncontrollable rush of pleasure between the belly button and the vagina" (*Women's Health*); "deep, all-encompassing, hair-pulling, eyes-in-the-back-of-your-head orgasms" (*Marie Claire*); "like being on Ecstasy" (*Salon*); "very intense feelings of love and spiritual transcendence" (*Cosmopolitan*); and "life-changing" (*Shape*).

One difficulty with such accounts is that in all likelihood, other structures, such as the vagina and possibly even the clitoris, were being stimulated along with the cervix during these so-called cervical orgasms. To focus specifically on the cervix, as well as bring some scientific objectivity to the topic, Whipple and her colleagues cobbled together what can only be described as a miniature toilet plunger. It consisted of a contraceptive diaphragm Velcro'ed to the tip of a tampon casing, and the other end of the tampon was connected to a purpose-designed holder that included a force transducer.[10] Volunteers, some of whom had spinal cord injuries, were shown how to insert the plunger so that the diaphragm locked onto the cervix, forming an airtight seal. Then they were encouraged to stimulate themselves by pushing and pulling on the plunger, but with strict instructions not to exceed the modest force of 180 grams (6.3 ounces). Evidently the researchers were concerned that a volunteer might injure herself in the quest for an elusive cervical orgasm.

Several of the volunteers, including some with spinal cord injuries, did experience orgasm when they stimulated their cervixes. The orgasms were documented by both self-report and recording of the typical increases in blood pressure and pulse rate. One of the cord-injured volunteers could not feel the stimulation due to her injury, but nevertheless she did experience an orgasm. This suggested that sensory information from the cervix sufficient to trigger orgasm can travel to the brain by some pathway that by-passes the spinal injury. This might involve the vagus nerve, which carries certain kinds of sensory information from the viscera directly to the brain stem.

Unfortunately, the researchers did not mention anything about the quality or intensity of the orgasms experienced by their volunteers. It could be that the orgasms were less than "life-changing"—perhaps on account of the awkward experimental setup and the crude pumping action of the plunger. Contrast this with the delicate touch described in an account from a Reddit thread: "She inserts her index and middle finger, places them on either side of my cervix, and gently rotates her wrist. And I completely lose control of my body, I'm all hers."[11]

Some women, like most men, ejaculate at orgasm, even though there is no obvious reason why they should do so. There are two kinds of female ejaculation. One is the forceful discharge of a large volume of clear fluid from the urethra—a phenomenon colloquially known as "squirting." The other is the release of a much smaller volume—perhaps a teaspoonful or so—of a slippery, cloudy fluid that exits the urethra without any great force.

Some women report one kind, some report the other, some report both, and many women report neither.

How many women ejaculate? Masters and Johnson claimed to have witnessed at least seventy-five hundred complete female sexual response cycles without ever seeing either kind of ejaculation. According to the International Society for Sexual Medicine, on the other hand, between 10 percent and 50 percent of women say they ejaculate, and even those who say they don't ejaculate may do so retrogradely, into the bladder.[12] Based on these estimates, we can state with confidence that the number of women who ejaculate lies somewhere between zero and 100 percent.

Where do these ejaculated fluids come from? The definitive answer was provided in the mid-1990s by the sex educator Gary Schubach. At the time he was a graduate student at the Institute for Advanced Study of Human Sexuality, a now-defunct institution based in San Francisco.[13] Schubach recruited seven "highly trained" women, all of whom said that they expelled fluid during orgasm. In a clinical examination room, the women stimulated themselves manually or with the help of their partners until they were near orgasm. At that point they were interrupted by a doctor, who outfitted them with a Foley catheter—a tube passed up into the bladder, where a balloon was inflated to prevent it from being dislodged. The women emptied their bladders though the catheter. Undaunted by this medical interruption, the women resumed their sexual stimulation, with the catheter in place, until they reached orgasm.

Some of the women experienced both kinds of ejaculation and some just one. In the case of women who experienced a large-volume discharge, the fluid exited via the inside of the catheter, demonstrating that it originated in the bladder. It was urine—presumably, urine that was left in the bladder after the emptying or that had entered the bladder from the kidneys after it was emptied. By calling it urine I don't mean to dismiss it, however. It was a sexual ejaculation, just not the same kind that men experience.

The small-volume ejaculate exited the urethra via the outside of the catheter. In other words, it originated from a source that discharged into the urethra somewhere between the bladder and the urethral opening. There's only one known candidate for such a source: the paraurethral glands, which, in developmental terms, may correspond to the male prostate. These glands are located adjacent to the urethra, and their ducts deliver the glands' secretions into the lumen of the urethra. One component of the paraurethral secretions is PROSTATE-SPECIFIC ANTIGEN (PSA), which in males is secreted exclusively by the prostate gland. PSA is present in the small-volume female ejaculate, just as it is in male ejaculate. Given these parallels, the paraurethral glands are sometime referred to as the "female prostate."

Schubach's work, which has been confirmed by others, showed that the large-volume female ejaculate comes from the bladder, whereas the small-volume ejaculate comes from the paraurethral glands. It did not show what the function of either ejaculate, if any, might be. It's not likely that the large-volume ejaculate has any function beyond possibly being a source of pleasure to the woman or her partner. Various suggestions have been made about the paraurethral ejaculate; for example, that it might help lubricate the vagina during intercourse. Given that this ejaculate contains PSA, it's possible that it plays the same role as does the PSA in men's ejaculate (see the following section).

Is the subjective experience of orgasm the same for women and men? Most of us only get to experience male orgasm or female orgasm, not both, so we can't easily make comparisons. But the ancient Greek seer Tiresias did experience both. According to legend, he was transformed into a woman as "punishment" for some minor transgression; seven years later, he was changed back to a man, whereupon he stated that sex was ten times better for women than for men.

People who transition these days—by medical intervention rather than divine wand-waving—don't generally agree with Tiresias, but they do note some differences. According to a Belgian study, those who transition from female to male report that their orgasms become shorter but more intense, whereas those who transition in the other direction say their orgasms become smoother and longer.[14]

Of course, many factors may cause the orgasms of transsexual individuals to differ from those of other people, especially the technical limitations of the transitioning process, as well as the psychosocial issues that may affect the sex lives of transgender individuals both before and after transition. Thus some researchers have addressed the question of subjective sex differences simply by asking cisgender (nontransgender) women and men to describe their orgasms. In one much-cited study from the 1970s, professional psychologists and others were unable to distinguish between descriptions of orgasms written by women and men.[15] In a more recent study, the Canadian psychologists Kenneth Mah and Yitzchak Binik had more than sixteen hundred students rate the validity of sixty adjectives as descriptive of their orgasms. Only one—"shooting"—showed a substantial sex difference; it was rated more highly by men, presumably because of its association with ejaculation. All in all, it seems that orgasms feel pretty much the same regardless of a person's sex, but there may be subjective differences related to ejaculation in men that are not experienced by women—at least not by women who don't ejaculate.

The major difference between the sexual response cycles of women and men is that some women can experience multiple orgasms before going into the resolution phase of their response cycle, whereas most men must make do with just one. Among women who say they are multiorgasmic, the number of orgasms can range up to one hundred, but the median number is just three, according to a survey published in 2021.[16] The orgasms may occur during a period of nonstop physical stimulation or with brief pauses.

The sex difference in the capacity for multiple orgasms is not absolute. Many women never experience multiple orgasms, and a few men do experience them naturally or can learn to do so. In such men, there may be a series of orgasms without ejaculation, followed by a final orgasm with ejaculation. However, a few men can experience a sequence of ejaculatory orgasms within a single response cycle. Masters and Johnson witnessed at least one demonstration of multiple ejaculatory orgasms. In 1998, Whipple and her colleagues also documented the phenomenon: their volunteer experienced six ejaculatory orgasms in the laboratory over the space of thirty-six minutes without losing his erection, and he desisted only because "the room became too hot and stuffy to continue." All six ejaculates were tested and shown to be semen.[17] These days it's easy to find NSFW videos online that illustrate the phenomenon.

Why do women and men generally differ in their capacity for multiple orgasms? An intriguing possibility is that it has to do with a hormone named PROLACTIN, which is secreted by the pituitary gland. Prolactin's best-known function is to promote breast development during a woman's pregnancy and milk production after childbirth. It has other functions in both sexes, however. In particular, it has an inhibitory effect on sexual feelings and behaviors. The pituitary gland releases a surge of prolactin into the bloodstream at the moment of orgasm, and this surge appears to act as a damper on subsequent sexual responsiveness.

A German group conducted an endocrinological study of a man who was capable of multiple ejaculatory orgasms. He turned out to be very unusual in hormonal terms: he experienced no orgasm-related rise in prolactin levels whatsoever.[18] Was that the reason he was capable of multiple ejaculatory orgasms? Findings in a single man are suggestive but not conclusive, so the study needs to be repeated in more multiorgasmic men.

The same research group tested the effects of lowering prolactin levels in regular single-orgasmic young men[19] using the drug CABERGOLINE, which reduces the secretion of prolactin. While under the influence of this drug, the men experienced an intensification of almost all aspects of their sexual responses. In addition, the researchers tested the men's ability to experience a second response cycle, with orgasm and ejaculation, thirty minutes after the first one. It may be that they expected only the cabergoline-treated men to achieve a second orgasm. In fact, however, all the men could do so, drugged or not—these were healthy young college students, after all. In such men the "refractory period" is very short.

Two orgasms separated by thirty minutes of "down time" don't qualify as multiple orgasms, which are orgasms occurring within a single response cycle. Thus the study didn't answer the question of whether lowering prolactin levels would allow single-orgasmic men (or women) to become multiorgasmic. However, cabergoline does help men who have difficulty experiencing any orgasms to become single-orgasmic.[20]

Anal sex—penetration of the anus by a penis, finger, tongue, strap-on dildo, or vibrator—has received much less attention from sexologists than vaginal sex. It was not mentioned in Masters and Johnson's *Human Sexual Response*, for example, and it earned only the briefest of mentions in Whipple's and colleagues'

Science of Orgasm. Yet anal sex is far from a rare sexual practice. In a large national survey published in Britain in 2019, 13 percent of adult women and 3.6 percent of men said that they had been on the receiving end of anal sex within the previous twelve months. As for the insertive role, 12.5 percent of heterosexual men, 52.4 percent of gay men, and 30.6 percent of bisexual men had taken that role within the same period.[21]

It's not difficult to understand how men reach orgasm through insertive anal sex. In fact, even if a vagina is available, some heterosexual men may find it easier to reach orgasm through anal penetration because of the greater tightness of the anus compared with the vagina. But what about the woman or man who takes the receptive role?

In the case of men, there's the prostate gland, of course, whose rear surface lies adjacent to the front of the rectum and is likely stimulated during anal penetration. No one has carried out a scientific study of the prostate during anal sex, as far as I'm aware, but according to published accounts, orgasms triggered by anal penetration with a penis or by use of a specialized prostate massager are often very intense "Super-Os" that involve much of the body rather than just the genital area.[22]

These orgasms may be accompanied not by ejaculation but only passive leakage of a small amount of prostatic fluid. In this case, the man may not enter a refractory period and instead may be capable of experiencing multiple prostate orgasms before finally ejaculating through other means. The penis may not even be erect during prostate orgasms. It would be interesting to know whether prostate orgasms without ejaculation are accompanied by the prolactin surge that accompanies regular ejaculatory orgasms.

Descriptions of men's prostate orgasms are similar to some women's descriptions of G-spot orgasms. If those orgasms result

from stimulation of the paraurethral glands, this functional similarity bolsters the idea that the female paraurethral glands and the
male prostate gland are developmentally equivalent structures.
Nevertheless, anal penetration in women is unlikely to provide
much stimulation of the paraurethral glands, because the vagina
is interposed between the anus and the glands. Women who
experience orgasm during anal sex presumably do so on account
of direct stimulation of the anus and the skin around it. Another
possibility is that anal penetration may trigger a cervical orgasm.

THE SECRET LIFE OF SEMEN

If you ask people where semen—men's ejaculate—comes from,
many will say the testicles. They're wrong, only about 1 percent
of the volume of the semen comes from that source, although
it's an important 1 percent, as it contains the sperm cells, or
spermatozoa—all 100 million or more of them. People with
more biological education will say "A little from the testicles and
the rest from the prostate gland." They're wrong, too. The bulk of
the semen—70 percent of it—comes from the most important
glands that no one has ever heard of: the SEMINAL VESICLES.
These glands are perched behind and above the prostate gland,
one on each side of the midline. Urological surgeons are so con-
fident that you've never heard of them that when they remove
your prostate gland (should you be so unfortunate as to require
that procedure) they remove your seminal vesicles, too, usually
without even mentioning the fact. You won't miss them, they
assume, because you didn't know you ever had them.

At the beginning of orgasm, the seminal vesicles, the prostate gland, and the left and right VAS DEFERENS (the tubes that
convey sperm from the testicles) all dispense frugal helpings of

their secretions into the rear portion of the urethra, which is enwrapped by the prostate. This loading process is called SEMINAL EMISSION. The secretions remain at the back of the urethra for just a second or two, and then spasmodic contractions of pelvic floor muscles expel everything from the penis in a series of pulses—ejaculation. Some reloading of the urethra presumably occurs between the pulses, although I haven't seen any experimental evidence for that.

In 2000, five male urologists at the University of Bern in Switzerland (the school known for its smelly T-shirt studies, mentioned in chapter 2) thought that it would be interesting to record the pressure inside the urethra during ejaculation.[23] To do this, they needed to place a specially designed probe in the urethra, an invasive procedure that would ordinarily involve obtaining approval from their institution's Human Use Committee. In order to circumvent this regulatory burden, or possibly for lack of willing volunteers, the urologists decided to perform the experiments on themselves.

In all five men, the highest pressure was recorded at the sphincter between the bladder and the prostate gland. This pressure, equivalent to that exerted by 5 meters of water, was maintained throughout orgasm. It was caused by the tight contraction of the sphincter, which prevented urine from mixing with the semen or semen from being driven backward into the bladder. Where the urethra passes through the prostate gland, the researchers measured pulsatile pressures of up to 3.5 meters of water, or 5 pounds per square inch, the pressure that is exerted directly on the semen and expels it from the penis. In theory, such a pressure would allow a man's ejaculate, if directed upward, to strike the bedroom ceiling. In reality, however, the ejaculate is greatly slowed by frictional losses and its own viscosity as it passes along the urethra.

Masters and Johnson never attempted to measure the height reached by a man's ejaculate, but they, as well as Alfred Kinsey before them, did take note of the horizontal distance over which it can be propelled. They found—as have generations of schoolboys—a great deal of variability: in some individuals, the ejaculate is flung several feet; in others, it just dribbles out. As far as is known, this variability is irrelevant to a man's ability to impregnate a woman. In the case of two men having intercourse with the same woman within a short period, however, the more forcefully ejected semen might get closer to the cervix or displace the other man's ejaculate. This would be a form of "sperm competition," a phenomenon that is known to occur in some species and may also have been relevant during human evolution. According to the psychologist Gordon Gallup of University at Albany, SUNY, for example, evolution has shaped the human penis into a form adapted to scoop out semen previously deposited in a woman's vagina by another man.[24] Short of rerunning the tape of evolution, it's hard to assess the merit of this idea.

The prostate gland and the seminal vesicles secrete two different sets of chemicals that must be kept separate until ejaculation, because they react with each other. In this they resemble the two defensive glands of bombardier beetles, whose secretions, once mixed, turn into a boiling and caustic spray. The male ejaculate is neither boiling nor caustic, fortunately, but it is still the site of some remarkable biochemistry.

There are four major actors in this drama. Two of them come from the prostate gland: first, negatively charge ions of the metal zinc, and, second, prostate-specific antigen (or PSA), an enzyme that cuts up proteins. (It is also the focus of the anxiety-provoking PSA test.) From the seminal vesicles comes a protein known as SEMENOGELIN. From the left and right vas deferens come the sperm. (More precisely, each vas deferens joins the duct of a seminal vesicle, forming a shared tube called the "ejaculatory duct," which leads into the urethra.)

These four constituents mix during or immediately after ejaculation. When semenogelin contacts the sperm, it immobilizes them. Zinc ions from the prostate gland then cross-link the semenogelin molecules to each other, causing the semen to coagulate into a tapioca-like gel. (If you were not fed tapioca as a child, you may be familiar with it as the little gelatinous balls in boba tea.) The sperm are now trapped in this network.

If semen is deposited in a woman's vagina, coagulation may serve the function of reducing the loss of semen by leakage from the vagina. In addition, though, the trapped spermatozoa now begin a process called "capacitation," a biochemical maturation that facilitates rapid swimming and recognition of the ovum as their target. This process is not complete until later, however, when sperm have reached the woman's oviducts (fallopian tubes).

At the same time that the semen coagulates, the PSA in it begins to attack and break down the semenogelin. (Conceivably, PSA from a woman's paraurethral gland ejaculation could contribute to this process if her ejaculate is carried into the vagina.) By ten minutes or so after coagulation, the semen is sufficiently liquid that spermatozoa are able to escape the network and, if they are in a woman's vagina, begin their migration up her reproductive tract. The partially dissolved coagulum may act as a kind of quality filter, allowing the healthiest spermatozoa to escape and holding back those with defects, of which there are always a large number. In fact, according to some analyses, most spermatozoa in a "normal" ejaculate are abnormal in some way. It seems that Nature has not yet perfected the process by which sperm are manufactured, or else she simply prefers quantity to quality.

Of course, ejaculated semen doesn't always find itself inside a woman's vagina. If a man ejaculates into water, as in a shower or bath, he may observe that the semen forms a stringy material. This is probably because the PSA diffuses away or loses its activity in the watery environment, so the semenogelin is not cut up and coagulation continues further than it otherwise would. In some animals, especially rodents, coagulation proceeds so far that the semen forms a solid plug in the female's vagina; this prevents any subsequent male from inseminating her—another example of sperm competition.

Without getting too involved in reproductive biology, it's worth mentioning one other aspect of the process that leads up to fertilization, because it's relevant to the issue of infertility. It has long been suspected that the ovum secretes a substance that attracts sperm, and in 2003, a group based at Ruhr University Bochum in Germany provided evidence that the attractant is a substance named BOURGEONAL, or a closely related molecule.[25] Bourgeonal is an organic liquid whose odor resembles that of lily-of-the-valley flowers; for that reason, it is used in the manufacture of perfumes. Bourgeonal is the only known odorant to which men are more sensitive than women; they can detect bourgeonal vapor at about half the concentration that women can.

Men's performance pales beside that of male mice, however, who can detect bourgeonal vapor at one part in ten quadrillion.[26] This ratio is approximately the same as that between a single drop and the volume of two hundred Olympics-sized swimming pools. Why a mouse's nose should be so sensitive to bourgeonal is not entirely clear, given that it's the sperm's sensitivity that's important for finding an ovum, not that of the nose. One possibility is that a male mouse's nose is capable of detecting whatever tiny amounts of bourgeonal (or of the related molecule, if bourgeonal itself is not the natural attractant) may spread from a female's ovum to her vaginal secretions, in which case the male would have direct information about the presence of an ovum before even mounting the female. This possibility hasn't been investigated, as far as I'm aware.

More recent studies have found that men who are infertile for unknown reasons are not as sensitive to bourgeonal as other men, their sperm are less strongly attracted to this substance, and they tend to possess different variants of the gene for the bourgeonal receptor from those possessed by fertile men.[27] These findings may help such men to become fathers: their sperm, though poor performers in bourgeonal sensitivity tests, should be perfectly competent to fertilize ova in in vitro fertilization procedures, where the sperm are placed in direct proximity to an ovum or actually injected into it.

WHEN DESIRE IS LACKING

I wrote earlier that people enjoy sex, but that's not always true. Sex can be painful, performance can disappoint, desire can be lacking. Sometimes medical science comes to the rescue; sometimes talk is more helpful. With some sexual problems, experts argue about which approach is best. Sometimes nothing helps.

I'm going to discuss just one of these problems: a woman's lack of sexual desire, a weak or absent sex drive. I'm picking this topic for the very reason that it's so challenging to think about in scientific terms. First, not wanting sex is not a problem unless a woman wants to want to have sex—unless she desires desire. So there is a double layer of subjectivity, which removes the topic from the objective world within which science operates most comfortably. When a woman does desire desire, her problem has been given a diagnostic name: HYPOACTIVE SEXUAL DESIRE DISORDER, or HSDD. HSDD has now been folded into a more general "female sexual interest/arousal disorder," but HSDD is still widely considered to be an appropriate diagnostic category.

A second issue regarding lack of sexual desire is that cultural expectations regarding it—especially in women—have varied greatly over time and place. Often women haven't been expected to initiate sex but only to respond to men's advances, in which case, feeling or expressing sexual desire may be seen as more pathological than a lack of such feelings. In such cultures it has only been a woman's failure to respond to men's advances— "frigidity," as it was traditionally called—that has been seen as a condition requiring treatment.

Conversely, there are other cultures—and this is increasingly true of contemporary America—in which women are expected to have an active sex drive, and a woman who says she's too busy or too old to think about dating or sex may be seen as deficient. Such cultural expectations are bound to influence a woman's own sense of having or not having a problem that needs remediation.

In some models of human sexuality, such as one put forward by the sex therapist Helen Singer Kaplan in the 1970s, sexual desire is the state that normally precedes and permits sexual arousal.[28] So we could think of it as mental arousal that is not yet accompanied by the physiological signs of arousal and may not yet be focused on a particular target. The colloquialism "horny" describes it aptly. We usually think of that term in reference to men—for example, a man who is prowling the clubs on Saturday night in an ever-more-desperate search for a willing sex partner. But women can be horny, too, of course.

Some people never desire partnered sex over their entire lifetime because they don't experience sexual attraction to anyone. Many of these asexual people do masturbate, but while doing so they don't experience fantasies of sex with a partner. Asexuality should probably be thought of as a sexual orientation—one that lies at the zero crossing of heterosexual and homosexual attraction—and it's not especially uncommon: estimates of its prevalence range from 0.5 to 3 percent of the population, with

1 percent being the most commonly cited number.[29] Most asexual people are satisfied with their orientation, which can have advantages, such as leaving more time for careers and other activities. There can also be difficulties, such as having to deal with expectations of sex by romantic partners. Still, asexuality is not a problematic lack of sexual desire, and it therefore doesn't qualify as HSDD.

Although there are women who remain sexually active into old age, many lose interest in sex with advancing years, and this loss may be viewed positively. "My lack of sex drive has been enormously liberating," one woman told Britain's *Guardian* newspaper. "I look back with some regret at the years I wasted on men."[30] So far from being a disorder, this woman's loss of interest in sex might be more appropriately labeled a recovery.

Yet there are also women of all ages who have little or no sex drive but would like to have one. They may feel they are missing something that their peers enjoy. They may remember with pleasure the sex drive they once had. And perhaps most commonly, they might wish they were turned on by their partner because that partner desires and expects sex with them. Here's how one twenty-one-year-old woman put it, as cited by Meston and Buss: "Having been with my previous boyfriend for three years, our sex life declined due to my disinterest. At times his discontent with the situation was so overwhelming and disruptive to the rest of our lives that I would feign interest and have sex with him just to make him happy, in part because I felt I wasn't holding up my end of the relationship sexually."[31]

This woman's lack of sex drive may have been specific to the particular relationship she was in: it may have been a consequence of dissatisfaction with nonsexual aspects of the relationship, in which case couples therapy or termination of the relationship might have solved the problem. But there are also women who lose interest in sex with anyone—who would not be interested even if *The Bachelor* came calling—and yet they want to regain it. These are the women whose condition is most appropriately referred by the term HSDD.

Sometimes HSDD can be treated by addressing its cause or causes, which may include excessive weight loss, painful intercourse, menopause, depression, overwork, or adverse social circumstances. According to some therapists, feminists, educators, and writers—Leonore Tiefer of New York has been a leading example of all four—a woman's lack of interest in sex can be understood and alleviated only by considering the entire social environment in which she lives. Thus Tiefer and others have campaigned against an approach that involves the "invention" of medical disorders such as HSDD and the quest for drugs to treat them, an approach that Tiefer has condemned as "disease-mongering."[32]

Yet the sex drive does have biological underpinnings. Two sex hormones in women, estrogen and testosterone (along with some related compounds) are the major players. Estrogen is probably more important than testosterone. Women's sex drive increases greatly at puberty and decreases during and after menopause, at least in part because of the corresponding rise and fall in the secretion of estrogen by the ovaries.[33] Yet sexual desire may persist to a variable degree long after menopause, either because of sex hormones that come from a different source—the adrenal glands—or because women's sexual desire is partly independent of sex hormone levels.

In postmenopausal women who complain of low sexual desire, treatment with estrogen is generally effective in restoring desire. This may be in part because estrogen relieves the vaginal dryness that can make intercourse painful, thus making the idea of having sex more appealing, but estrogen also acts directly on the brain to increase desire.

Since the beginning of this century, pharmaceutical companies have promoted the idea that testosterone is as important or perhaps more important than estrogen in maintaining women's capacity to experience sexual desire. According to a review of controlled trials by Maurand Cappelletti and Kim Wallen of Emory University, testosterone does exert some effect of that kind, but only when given in combination with estrogen and in doses that raise testosterone levels far above those normally found in women.[34] Such high doses risk causing unwanted side-effects, such as the growth of facial hair. Thus the reviewers recommended treatment with estrogen only. The Food and Drug Administration has not approved any testosterone treatment for women, but that has not prevented women's doctors from writing millions of off-label prescriptions for the hormone. In this context, Tiefer's critical stance toward the pharmaceutical and medical industries may be justified.

Ever since 1998, when Viagra (sildenafil) exploded onto the scene and revolutionized the treatment of male erectile dysfunction, the search for some kind of female equivalent has been under way. In fact, some feminists have taken a line diametrically opposite to that taken by Tiefer: they have complained that women's sexual problems, including a lack of desire, have been ignored by pharmaceutical companies.

Neither Viagra nor other drugs in the same class alleviate low sexual desire in women, which is hardly surprising as they do not have that effect in men either. Regarding sexual performance, Viagra has been shown to offer some relief in women whose sexual function has been negatively impacted by antidepressants. In particular, in a controlled trial, it has been shown to help women on antidepressants who have difficulty experiencing orgasm—a frequent complaint in such women.[35] Otherwise Viagra does not seem to alleviate women's sexual problems.

More recently two quite different drugs, FLIBANSERIN and BREMELANOTIDE, have been approved by the FDA for the treatment of low sexual desire in women who have not yet reached menopause. Flibanserin, marketed as Addyi, was approved in 2015. It works by raising the levels of dopamine and another neurotransmitter, norepinephrine, in the brain. Bremelanotide, marketed as Vyleesi, was approved in 2019. It has to be self-administered by injection forty-five minutes before an expected sexual encounter. Its mode of action is not fully understood.

There is an interesting story behind the development of bremelanotide. A chemical precursor of the drug was being investigated as a possible sunless tanning agent when one of the researchers, Mac Hadley, then at the University of Arizona College of Medicine, had a curious experience. In his own words,

> I tested the efficacy of the peptide to produce a tan on myself. . . . [It] caused a rather immediate, unexpected response: nausea and, to my great surprise, an erection. . . . While I lay in bed with an emesis [vomit] pan close by, I had an unrelenting erection (about 8 h duration) which could not be subdued even with a cold pack. When my wife came upon the scene, she proclaimed that I "must be crazy." In response, I raised my arm feebly into the air and answered, "I think we may become rich."[36]

So far from becoming rich, however, Hadley died in a tragic incident before bremelanotide came to market: he was murdered in his own home when he surprised a burglar.

Although the initial investigations into bremelanotide's sexual effects were carried out on men, the drug was eventually approved only for use by women, based on clinical trials sponsored by the manufacturer in which women with HSDD who took the drug reported greater sexual desire than those taking a placebo.[37]

Double-blind placebo-controlled trials, such as those that led to approval for flibanserin and bremelanotide, are about as scientific as you can get in clinical research. Even so, there is reason for skepticism as to whether these drugs offer a meaningful boost in sexual desire. First, the beneficial effects of the drugs, though statistically significant, are small in degree. In a 2016 meta-analysis of eight studies of flibanserin, women's ratings of their overall improvement were either no different from placebo or only minimally better.[38]

Second, the drugs have side effects. Flibanserin can cause dizziness and low blood pressure. Bremelanotide raises blood pressure and frequently causes nausea—not a feeling one welcomes while having sex. (In fact, nausea-inducing chemicals are sometimes used as conditioning agents in efforts to turn people off unwanted types of sexual behavior.) In the trials of bremelanotide that led to its approval by the FDA, far more women who took the drug dropped out of the trial than did women in the placebo group, presumably because of side effects such as nausea. Only 40 percent of women who took the drug completed the trial and opted to continue it afterward, whereas 69 percent of the women who took the placebo did so—a testimony to the power of suggestion in medical therapeutics. If women (or men) take bremelanotide too frequently or at too high a dosage, they risk developing irregular skin pigmentation as well as precancerous skin lesions.[39]

Another factor that reduces the persuasiveness of these and many other drug trials is the degree to which the trials have been influenced by the manufacturers of the drugs under investigation. In the case of the bremelanotide trials, all eight authors of the published study were employed by or received funds from the developer or licensee of the drug, the trials were paid for by the developer, and the licensee also hired an outside person to "provide editorial support in the preparation of the manuscript"—thought to be code for "wrote the paper." It could be for this reason that the numerical data cited in the previous paragraph, which spoke against the usefulness of the drug, were not included in the paper and had to be dug out of the original FDA reports by an independent party, the psychologist Glen Spielmans of Metropolitan State University.[40]

At the beginning of this book, I mentioned the 237 reasons why humans (or college students) have sex, according to the survey conducted by Cindy Meston and David Buss. Some of the proffered reasons, such as "I was attracted to the person" or "the person was too hot to resist," were fully consistent with the model put forward by Kaplan, according to which sexual desire is the precursor and motivator for sexual behavior. Yet many other reasons given by the students had no obvious connection with sexual desire. So how common is it that people enter into sexual encounters without prior erotic motivation?

The answer is very common, especially among partnered women, according to the psychiatrist Rosemary Basson of the University of British Columbia.[41] Basson believes that many women enter into sexual encounters because of a nonsexual interest in sex, a desire for intimacy, for example, or simply as a way to please the partner. But as the encounter begins, sexual stimulation—both physical and psychological—generates sexual desire. In other words, there is a circular, self-reinforcing relationship between sexual activity and the intensity of desire. This process may involve not just the activation of brain regions that drive sexual arousal but also the deactivation of brain regions that, most of the time, inhibit arousal.

Thus Basson encourages a woman who complains of low sexual desire to embark on sex with her partner when it seems appropriate, even if she is not conscious of a desire to do so. The only precondition: she should make sure that her partner understands how to give her sexual pleasure and satisfaction. The hope is that sexual desire will develop as the encounter proceeds, and a satisfying encounter increases the likelihood that the woman will desire the next one. "Rewarding sex provides its own incentive for its repetition," Basson and two colleagues wrote. "No 'desire drug' needed."[42]

Given the many potential causes of low sexual desire, it's not likely that a single approach to treatment, such as the one suggested by Basson, will work for all women. There is a place for a desire drug if an effective and safe one can be discovered.

CONCLUSIONS

I started this chapter by providing scientific evidence in support of an obvious fact: people want to have sex because having sex makes them happy. Part of the reason it does so is that while having sex, people are unusually "in the moment" and less easily distracted by wandering thoughts. I ended the chapter on a contrasting note: some people don't want to have sex—not partnered sex, at least—either because they've never been sexually attracted to anyone or they have lost their sexual motivation, a loss that may be experienced as a blessing or as a problem that may be alleviated by therapy or drug treatment. In between, I discussed the physical basis of the events that form the climax of sexual behavior, orgasm and (for most men and some women) ejaculation. Surprisingly, it's not yet completely clear which anatomical structures are capable of triggering orgasm and whether the stimulation of different structures triggers different kinds of orgasm.

Regarding male ejaculation, science has elucidated the details of what happens: the coordinated secretions of three glands before ejaculation and the chemical interactions among the components of these secretions during the minutes after ejaculation. These processes are a reminder that even if the great majority of our sexual acts are intended to be recreational rather than reproductive, our bodies don't yet know that. They are constantly preparing for those rare or nonexistent occasions when, whether desired or not, pregnancy is a possible outcome.

I've omitted a great deal. This may be the first broadly based book about sex that includes no advice about sexual positions, for example, or about how to avoid pregnancy and sexually transmitted infections. More important, this chapter treats sex in a vacuum, divorced from the relationships within which it may take place. The following chapter makes amends for that deficiency.

6

RELATIONSHIPS

Most people want a loving and enduring sexual relationship at some point in their lives. An enormous amount of research has gone into discovering how to pick and keep the right partner. Has it come up with the answer?

The contemporary Western ideal, based perhaps on fairy tales we read as children, is to lead with the heart: two people meet by chance, fall in love, and live happily ever after. Sometimes that's what happens, but most often, probably, it doesn't. You fall in love with someone but that person doesn't fall in love with you, or they do but they turn out to be a dog. You've embarrassed yourself, but you've also learned a lesson that may help you do better next time.

The opposite of this ideal is the arranged marriage: the parents or someone hired by them search for a suitable mate, they negotiate, and you get to meet and consent—or not—before the nuptials. Arranged marriages, considered on a worldwide basis, are far less likely than autonomous marriages to end in divorce, but that doesn't tell us how happy those marriages are. Cultures in which arranged marriages are common are those in which divorce is difficult to obtain and, if it is obtained, will likely be followed by a variety of negative consequences, especially for women. So couples whose marriages were arranged tend to stay married come what may. The concept of arranged marriage runs counter to the Western principle of individualism, but proponents of arranged marriages say that love comes after marriage rather than before.

In the modern world, it can be difficult to arrange marriages, even in immigrant communities, because young people are likely to establish sexual relationships on their own and may move in together and even have children long before getting married, if they marry at all. Many arranged marriages are now "love-cum-arranged" marriages, in which an already committed couple approaches their parents and asks them to arrange their marriage, giving the parents the opportunity to keep up the appearance of a traditional match—and to pay for the wedding.[1]

Most long-term relationships, whether marriages or not, start out neither as instant love matches nor parental arrangements but as something in between: warmer than an arrangement, cooler than a love match. Traditionally, Westerners have been introduced to their partners by friends or relatives, and this method had distinct advantages over simply running into someone in a laundromat: the potential partner has in a sense been preapproved, and for that reason they are more likely than a randomly encountered person to share one's interests and attitudes. Also, the couple starts out with a mutual friend—the nucleus of a social circle.

The sociologist Michael Rosenfeld of Stanford University and his colleagues have conducted nationally representative surveys to understand how couples meet.[2] They find a dramatic shift over time, as shown in figure 6.1. Starting in the mid-1990s, the internet has largely replaced other ways in which prospective heterosexual couples make contact. In particular, the numbers

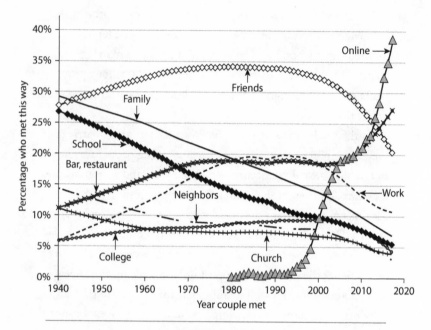

FIGURE 6.1. Where couples meet.

Source: From M. J. Rosenfeld, R. J. Thomas, and S. Hausen, "Disintermediating Your Friends: How Online Dating in the United States Displaces Other Ways of Meeting," *PNAS* 116 (2019): 17753–58.

of couples who meet through friends has taken a nosedive. The number who say they met in bars or restaurants has also risen, but that doesn't mean the singles bar has retained its popularity as a place to search for partners; rather, that's where many couples who make contact online arrange to have their first face-to-face encounter.

The main drawback to online dating is well known: you meet a persona, not a person, and that can lead to face-to-face encounters in which your only thought is how quickly you can get away. Experienced users of dating or matchmaking sites are well aware of the possibility of deception or incomplete disclosure. For that reason, unless they are seeking immediate sex, they engage in lengthy online communication before meeting. Despite such precautions, couples whose first contact was online are more likely to break up than comparable couples who met in other ways.[3]

An alternative way to vet a large number of potential partners is by participating in organized speed-dating events, in which you get to spend three to five minutes with each of a couple of dozen members of the other sex—or the same sex, if that's who you're looking for. At the end of the event, you list your "positives"; later, the mutual positives are given each others' contact information. The structured nature of speed dating makes it well suited to the study of how people find mates.

You may think that the opportunity for face-to-face conversation during speed dating would allow for a somewhat more rounded evaluation of potential partners than the appearance-focused judgments typical of online dating. But it doesn't, according to a study by Robert Kurzban and Jason Weeden, then at the University of Pennsylvania. Using data on more than ten thousand speed-dating participants provided by HurryDate (now defunct), those psychologists found that most participants still relied on the crucial traits of facial attractiveness, height, and body-mass index when making their decisions.[4]

Another informative study was carried out by Paul Eastwick and Eli Finkel of Northwestern University.[5] (Eastwick is now at the University of California, Davis). Female and male undergraduate students first answered questions about the characteristics they would look for when selecting partners at a speed-dating event. In agreement with numerous prior studies, women were more likely than men to emphasize the person's earning potential, whereas men were more likely than women to emphasize the person's looks. Yet when the participants later took part in a speed-dating event, the choices they made showed no such sex difference. In fact, stated preferences in general were very poorly predictive of actual choices. For example, a woman who listed earning potential as the most desirable characteristic might choose a man with poor financial prospects. Eastwick and Finkel interpreted their findings in terms of "choice blindness": the idea that people have only limited awareness of the reasons they make decisions in their lives.

Of course, like most other forms of face-to-face social interaction, speed dating was brought to a halt by the COVID-19 epidemic. Some speed-dating outfits moved online; whether that is a satisfactory format for making romantic contacts remains to be seen.

Although much relationship science has focused on choices that people make when they initially meet potential partners—as with dating apps, speed dating, and dates arranged by friends— many people establish romantic relationships with individuals they have already known for some time as nonsexual friends. According to a 2021 study led by Danu Stinson of the University of Victoria, British Columbia, about two-thirds of intimate relationships among college undergraduates and other young adults are of this "friends-first" type.[6] Friends-first relationships involving LGBT people are even more prevalent, perhaps because of the lesser size of the relevant communities. In an even smaller community, that of Deaf gay people, dating between strangers is actually rare because everyone already knows everyone, even if one includes as potential partners those hearing gay people who are fluent in American Sign Language.[7]

Friends-first relationships are not just common, they are also preferred by many young people. In Stinson's study, they were cited as the best route to a romantic relationship by nearly half of all college students, whereas online dating—the relationship juggernaut, according to Rosenfeld's analysis—was cited by only one of the 300 respondents. Of course, college students enjoy many more face-to-face contacts with potential partners than do most other adults, making the online avenue less necessary.

The advantages of knowing a person before getting romantically involved with them were illustrated in another study by Paul Eastwick, this time in collaboration with Lucy Hunt.[8] The participants were students in a large class that was broken into discussion sections of about seventeen students; the groups met weekly over the course of a semester, and all students were required to participate in the discussions. During the second session, the students completed questionnaires that asked about the value of their fellow students as potential mates. There was a fair degree of consensus about the relationship value of each student. When the questionnaire was repeated in the final week of the class, however, this consensus had largely broken down. It seemed that the initial judgments were based on physical appearance and common stereotypes, whereas the final judgments were based on personal qualities that were important to each person making the assessment—qualities that would be more likely to lead to successful romantic matching.

BIRDS OF A FEATHER? OR OPPOSITES ATTRACT?

When looking for love, do people desire and seek out partners like themselves? And if so, is that a good strategy? In other words, do relationships between similar partners offer the best prospects for lasting happiness? According to many studies, the answer to both questions is yes. But I'm skeptical.

Certainly couples resemble each other in many characteristics, and that phenomenon has a name: HOMOGAMY. (Sometimes this

term is used for married couples only, but I use it more broadly.)
The shared characteristics commonly include demographic factors such as race, age, educational level, socioeconomic class,
and religion; physical traits such as height, BMI, and facial
attractiveness; and intelligence, political attitudes, and, in some
studies, certain aspects of personality. These are statistical generalizations, of course—none of them applies to every couple—but
the weight of evidence demonstrating the similarity of couples is
overwhelming.

So why am I skeptical? One reason: the fact that a couple
is similar in some characteristics doesn't necessarily mean that
each person was specifically attracted to a partner resembling
themselves or that the similarity benefits their relationship. College students, for example, necessarily spend much of their time
with other college students, and for that reason alone, there's a
good chance they will form relationships with persons of similar
age and educational attainment, regardless of any intrinsic preferences. Similarly, religious observance brings together people of
the same religion. People live near members of the same race;
people of similar wealth shop at the same stores; and so on. You
have to go out of your way to meet people who are significantly
different from yourself, so the phenomenon of homogamy could,
in large part, be the consequence of sheer inertia.

When it comes to characteristics that have a broadly market-
able value, such as physical attractiveness, the operation of the
marketplace favors homogamy, regardless of people's uncon-
strained desires. Over the course of childhood, adolescence,
and young adulthood, people learn—sometime through bitter
experience—who they can "afford" as friends, lovers, or spouses.
Once ingrained with this self-knowledge, young people may seek
out partners at their own "price level," even though they may desire
more valuable partners. This marketplace model of relationship

formation was explored by the social psychologists Ellen Berscheid, Elaine Hatfield, and others in the 1960s and 1970s.[9]

There is, admittedly, some evidence that inertia and the marketplace are not the whole story. Regarding physical appearance,
for example, there have been studies in which each participant is asked to rate the attractiveness of computer-generated
opposite-sex faces, some of which have been manipulated to
partly resemble the participant.[10] This manipulation increases
the attractiveness of female faces as judged by men. Still, this
influence of physical self-resemblance is fairly weak—it is not
as strong as the influence of "objective" attractiveness; that is,
attractiveness as rated by a consensus of viewers. And there is
little, if any, comparable influence of self-resemblance when
women rate men's faces.

If people do seek sexual or romantic partners resembling
themselves, one might expect incest to be much more common
than it is. After all, close relatives such as siblings typically have
a great deal in common in both physical appearance and psychological and demographic characteristics. Siblings typically
grow up together, however, and close contact during childhood
exerts a kind of "negative imprinting" that works against sexual
attraction in adulthood. This so-called WESTERMARCK EFFECT,
named for the Finnish anthropologist who described it in the
late nineteenth century, is thought to be an evolved mechanism
that reduces the likelihood of incestuous matings.

The Westermarck effect does not operate between siblings
who are brought up separately, and such sibling pairs, if they
meet in adulthood, may indeed develop sexual and romantic
feelings for each other. Such cases make the news from time
to time, especially when two siblings, unaware of their related-
ness, partner up and have children. Although this phenomenon
has been given a pop-psychological moniker—"genetic sexual

attraction"—it's not clear that there is any greater likelihood of attraction or mating between siblings reared apart than there is between unrelated individuals.

In chapter 2, I discussed the possible influence of MHC (major histocompatibility complex) genes on sexual attraction. I mentioned that mice prefer to have sex with animals whose MHC genes are dissimilar to their own but that there's conflicting data on whether humans, too, are attracted to MHC-dissimilar partners. Even if they are, such chemical attraction has little or no impact on marriage choices: a recent large study found that the MHC genes of married couples are no more or less likely to be similar than one would expect on a chance basis.[11] In other words, if MHC preferences do play an "antihomogamy" role in partnership or marital choices, that role is weak and masked by the many other factors that come into play. However, once people are partnered up, dissimilarity in MHC genes (more specifically, HLA class I alleles) is associated with greater satisfaction with marital sex, greater satisfaction with the marriage as a whole, and a stronger desire to have children.[12]

There's one glaring exception to the generalization that people select partners resembling themselves, an exception that is rarely, if ever, mentioned in discussions of homogamy. This is the fact that most couples consist of two persons of different sexes. If homogamy rules, why isn't everyone gay?

The standard answer probably runs something like this: First, there must be an evolved mechanism to direct most people's sexual attraction to the other sex, otherwise our species would have gone extinct long before now. Second, heterosexual coupling doesn't conflict with homogamy, because gender differences are trivial if they exist at all, so a person can just as easily find a similar partner among members of the other sex as among members of the same sex.

The first point is likely correct; the second isn't. Despite a recent torrent of books and articles dismissing psychological gender differences as a sexist fiction, the fact is that the personalities and interests of men and women, and boys and girls, do differ from each other (on average) in a variety of significant ways. What's more, there's evidence that some of these differences are not simply imposed on individuals by cultural expectations, because they are most evident in societies in which women and men are most at liberty to do their own thing.[13] Therefore, the selection of an opposite-sex individual as a mate usually carries not just the obvious physical differences between partners but also a significant degree of psychological difference.

That issue doesn't arise for gay people, of course, or for bisexual people who choose same-sex partners; they are free to select near clones of themselves. Yet they don't. Same-sex couples are more likely than male-female couples to differ from each other in demographic characteristics such as age, race, and educational attainment. This emerges from the analysis of both census data[14] and social media. In a study by Facebook Data Science, for example, the average age gap between same-sex couples in their late thirties was twice as large as for opposite-sex couples in the same age range (eight versus four years).[15] The fifty-year age difference in the same-sex Wofford–Charlton marriage, which I wrote about in chapter 4, was certainly an outlier but one that many gay people would understand as being compatible with a genuine romantic relationship.

Regarding psychological characteristics in couples, numerical data are largely lacking, but there is a long tradition of gender-complementary same-sex relationships; that is, relationships in which one partner is more masculine in personality and the other more feminine. Among lesbians, butch-femme relationships have gone in and out of favor but have never disappeared entirely. Among gay men, the top/bottom distinction is based primarily on anal sex role (insertive or receptive), but it is also associated with gendered personality traits: tops self-report as more masculine than bottoms.[16]

I don't want to belabor these distinctions, because there are plenty of gay people and couples in same-sex relationships to which they don't apply. However, they exemplify an antihomogamy trend in same-sex relationships, which could be thought of as paralleling the "built-in" antihomogamy of heterosexual relationships.

So what effect does similarity or dissimilarity between couples have on their relationships? One could imagine either of two possibilities. Similarity might do a better job of stabilizing relationships because the partners would agree more; they would share attitudes and interests and thus reinforce each other's sense of self-worth. On the other hand, dissimilarity might do a better job because each partner would contribute something different—social versus analytical skills, for example—so that they complement each other. There is a kind of "deal," and that deal enables the couple to function better than a couple with more similarities.

In Aristotle's *Nicomachean Ethics*, the Greek philosopher discussed *philia*, the quality of a good relationship, sexual or otherwise.[17] (This is somewhat different from the current use of the term, which is often taken to exclude erotic relationships.) He disparaged "deal" relationships, comparing them with commercial deals that cease to function once the items traded can no longer be supplied or are no longer desired. A truly lasting relationship, he held, is one in which both partners are alike—especially in virtue—and can therefore love each other for their goodness regardless of life's vicissitudes.

In my view, Aristotle would have done better to stick with octopus sex, a topic on which he made some valuable observations. For humans, being of virtuous character is all well and good, but saintliness by itself is not a recipe for sexual satisfaction, and without sexual satisfaction, romantic relationships are liable to founder.

Contemporary studies have found no evidence that similarity in any dimension of personality is associated with greater satisfaction in marital relationships.[18] Nor, for the most part, have they found evidence for the opposite: that personality differences increase marital satisfaction. The published studies focus on male-female marriages, however, and pay little attention to gender characteristics.

Another shortcoming of research in this field is that most studies have been snapshots of young couples whose relationships have yet to be tested by time. This deficiency was remedied in an ambitious study by Michelle Shiota of Arizona State University and Robert Levenson of the University of California, Berkeley.[19] They recruited heterosexual married couples who were in their forties or older and had been married for at least fifteen years. At the start of the study, the participants underwent standard personality tests, which assess the "Big Five" factors in personality: extraversion, agreeableness, conscientiousness, openness to experience, and neuroticism. They were also assessed on measures of marital satisfaction. No clear relationship between similarity of spouses and satisfaction existed at that time. When marital satisfaction was assessed six and twelve years later, however, satisfaction was negatively correlated with similarity at time zero. In other words, couples who differed from each other in personality tended to become happier with their marriage over time compared with more similar couples.

Shiota and Levenson offered several possible reasons for their finding. One was the complementarity argument—the one rejected by Aristotle. Here's how Shiota and Levenson put it:

> For example, a couple in which one partner is achievement-driven and work-focused and produces a high income (a profile reasonably associated with Conscientiousness) and the other partner is more socially oriented, maintaining relationships outside the marriage as well as taking primary responsibility for raising the family (a profile reasonably associated with Extraversion and Agreeableness), may face less conflict in getting through a week's tasks than a couple in which both partners are workaholics or social butterflies. On a given evening, if someone needs to pay the bills and balance the checkbook, and someone needs to call other parents to arrange a carpool, the "complementary" couple will presumably argue less about who does what than the "similar" couple.

I believe this assessment is accurate but would go beyond it and add a less utilitarian, more romantic interpretation. Romantic love is a passionate desire for union with another person, but you can't desire something you already have. That's why in an established relationship, romantic love eventually fades, to be replaced, one hopes, by companionate love, which is a strong and durable liking, usually with an erotic element. To the extent that two people differ, to that same extent they remain strangers to one another, and that is what allows a spark of romantic love to persist—mystery rather than merging.

A final note on homogamy—from an economic perspective. Homogamy preserves group identities over the generations. This may be a good thing in some ways, but it also preserves the divide between rich and poor. In fact, now that so many couples are both breadwinners, homogamy increases economic inequality, because couples often consist of two high-earning or two low-earning partners.[20]

Women are far more focused than men on the incomes of their prospective mates.[21] You may think that this traditional pattern would lessen as women's own financial prospects improve, but in fact, highly educated women still seek to marry up the economic scale.[22] Thus social policies that seek to remedy economic inequality need to take account of the way in which homogamy and the trend toward dual-income relationships act to maintain that inequality.

COMMUNICATION

It's obvious that communication between partners or spouses is vital for the success of relationships, especially over the long term. If partners never talked, all kinds of bad things would surely happen, such as bills getting paid twice or not getting paid at all, and much worse. But what kind of communication, at what times, and on what topics?

Among relationship experts, the best known to the public may be the psychologist John Mordecai Gottman, who has studied couple relationships—specifically, marriages—for more than forty years. With his wife, Julie Schwartz Gottman, and others he runs the Gottman Institute, a Seattle-based organization that offers all kinds of programs and events designed to improve intimate relationships, especially those that have run into difficulties. Gottman bills himself as "The Country's Foremost Relationship Expert," and the popularity of his books, such as the 1999 bestseller *The Seven Principles for Making Marriage Work* (with Nan Silver), seems to bear that out.

Unlike most of those who dispense relationship advice, Gottman bases his ideas in science: specifically, studies of actual relationships in the "Love Lab" at the University of Washington, where he is now an emeritus professor. In a typical study, Gottman and his colleagues recruited couples early in their marital lives and videotaped their interactions while they discussed potential problems in their relationship. The researchers scored the videotapes for various kinds of positive and negative statements or comments by each partner, and they sometimes measured physiological parameters such as heart rate. Some years later, they checked whether the couples were still married or had divorced.

From these data Gottman calculated what kinds of interactions early in a marriage were associated with the likelihood that the couple would be divorced at a later time. In one much-cited 2000 study by Gottman and Robert Levenson, the researchers reported that expressions of negative feelings early in the marriage were predictive of divorce within seven years.[23] The most important such negative expressions were the wife's criticism, contempt, and defensiveness and the husband's defensiveness, contempt, and stonewalling. A particularly toxic style of interaction was DEMAND–WITHDRAWAL: a pattern in which one partner—usually the wife—criticizes or nags, and the other partner—usually the husband—clams up, buries himself in a crossword, or leaves the room. Couples who divorced later—after more than seven years—were characterized not so much by their negative interactions at the beginning of their marriage as by the absence of positive interactions at that early time.

Gottman has not been shy about his ability to forecast the future, either in his popular utterances or his research publications. "Through groundbreaking mathematical models integrating nonlinear differential equations, we can reliably predict and chart the future course of a relationship," announces the Love Lab's website.[24] "I can predict with great precision whether a couple will stay happily together or lose their way after listening to them interact for as little as fifteen minutes," writes Gottman in *The Seven Principles*.[25] "The model predicted divorce with 93 percent accuracy," the Gottman–Levenson study concluded.

Those claims may be misleading. "Prediction," in common parlance, means forecasting the future, but the model used in the Gottman–Levenson study was created retrospectively, after the study was complete. The study did not include a test of the model's ability to predict the marital outcomes for a new set of couples. What's more, Gottman and Levenson selected the couples to be used in the model in such a way as to make the model more predictive than it might otherwise have been. They did that by dropping from the set many "middling" couples, those whose future would have been difficult to predict because, early in their marriages, they were neither unusually satisfied nor unusually dissatisfied with their relationships. This was like developing a model to predict bankruptcy after removing participants who expressed a middling degree of satisfaction with their finances. The researchers also helped their case by taking into account any mention by their participants of a wish or intent to split up. It doesn't take scientific analysis to guess that such participants have a greater-than-average likelihood of divorce. Finally, the 93 percent claim may itself have been misleading: only 28 percent of the couples divorced, so a model that simply predicted that everyone stayed together would have achieved a 72 percent predictive accuracy. That's less than 93 percent, certainly, but not as much less as a guileless reader might imagine.

Gottman went on to apply his findings to help troubled couples improve their relationships, and that's what *The Seven Principles* is about. The principles are the following: learn about each other, focus on positive qualities, interact frequently, let your partner influence you, solve your solvable problems, overcome gridlock, and create shared meaning. Stated in this abbreviated fashion, the advice may sound a bit shopworn, but, of course, the therapy is in the details. Thus, for example, while encouraging couples to solve their solvable problems, Gottman lays out examples of problems that are not solvable: essentially "irreconcilable differences" that, without help, might trigger divorce. Gottman argues that the way to deal with such differences is not to struggle futilely to resolve them but, rather, to develop a nurturing atmosphere through frequent positive interactions. Such an atmosphere makes unresolvable problems less salient and therefore less threatening to the relationship.

I don't doubt the value of the precepts laid out by Gottman. Still, there is something of a chicken-or-egg question: are happy relationships the result of good communication, or is good communication the result of happy relationships? To address this question, a group led by Justin Lavner of the University of Georgia assessed the marital satisfaction and communication skills of couples at regular nine-month intervals after marriage.[26] They found very little tendency for good communication early in the marriage to predict greater marital satisfaction later.

In a longitudinal study of more than three thousand heterosexual couples published in 2021, a group led by Matthew Johnson of the University of Alberta reported that satisfaction and communication varied together over the course of a marriage, presumably reflecting changes in the couples' circumstances. But, again, changes in the amount or style of communication at a given time were not followed by predictable changes in marital satisfaction.[27]

Gottman has acknowledged that there is something deeper to a successful relationship than following behavioral precepts. "The foundation of my approach is to strengthen the friendship that is at the heart of any marriage," he writes in *The Seven Prin-ciples*. This underlying friendship is what makes his precepts pos-sible to follow.

How successful is Gottman's form of marital therapy? Much better than traditional methods, according to Gottman himself. He claims that among troubled couples who attended his work-shops, only 20 percent "relapsed," whereas the nationwide relapse rate for standard therapy, according to Gottman, is 30 to 50 per-cent. This claim can't be taken at face value, however, because it's not based on any objective comparison of his and other thera-pists' methods, or on any precise definition of a "relapse."

A couples therapy program designed by Gottman, along with two programs developed by other therapists, were tested in a multicenter study funded by the U.S. Department of Health and Human Services. That study, titled the Building Strong Fami-lies Project, was aimed at unwed couples who were expecting a baby or had just had one. At each site, couples were randomly assigned to the therapy program or to an untreated control group. According to Gottman, the findings of the study demon-strated the effectiveness of his program.[28]

The reality was different, however, according to the official report of the study. At none of the sites that used Gottman's pro-gram were the treated couples more likely to stay together or get married than the control couples, and at one site, they were significantly less likely to stay together. The only site at which there was a positive treatment effect on couples' stability used a program developed by a different therapist.[29]

None of this should be taken to mean that Gottman's ideas are wrong or that his form of relationship therapy is useless. It has always proven difficult to demonstrate the efficacy of psy-chotherapy of any kind using scientifically valid methods. That's probably because in the real world, some people will benefit greatly from therapy, some will not, and some will lose interest or drop out before a beneficial effect has a chance to show itself. Many other powerful factors act on couples' lives and relationships aside from any therapy program in which they may participate, and so the effects of therapy may get lost in the noise. And, most likely, the theory underlying any particular treatment program is less important than the humanity of the therapist and the level of trust that can be established between the therapist and the troubled couple.

THE GREEN-EYED MONSTER

In 1992, the evolutionary psychologist David Buss, along with three colleagues, published a study on the topic of jealousy in intimate relationships.[30] The paper reported a difference in the way that women and men experience jealousy. Men were more likely than women to experience jealousy triggered by the belief that their partner was having sex with a third party. I'll call that SEXUAL JEALOUSY. Women, on the other hand, were more likely than men to feel jealous triggered by the belief that their partner was romantically entangled with a third party. I'll call that EMOTIONAL JEALOUSY. This sex difference was far from absolute—both sexes experienced both kinds of jealousy to some degree—but it emerged robustly in several experimental studies.

Buss interpreted these findings in evolutionary terms. Men cannot be certain who their biological children are—or could not be certain before the invention of DNA testing. Thus they have always faced the risk that they might parent a child fathered by some other man. Women have always known who their biological children are, but they have faced a different risk: the risk of abandonment by their partners in favor of another woman, with loss of their partners' investment in parenting. In both sexes, jealousy evolved to mitigate those risks—by triggering such behaviors as mate guarding, physical or psychological aggression toward rivals, or enhanced solicitude toward partners when rivals threaten—but men's and women's jealousy tends to be triggered more strongly by threats to the reproductive interests of each sex.

Buss's 1992 paper became something of a citation "classic." It has spurred many other studies, by Buss and others, that have tested or extended his conclusions in a variety of ways. Some studies have challenged Buss's findings or his evo-psych interpretation of them, but many others have strengthened his conclusions. In 2018, Buss published a retrospective review that had the flavor of a self-congratulatory "victory lap:"

> In short, efforts to explain [our] findings with alternative theories, and to explain away the findings as being due to a perceived problem with one method, have failed. Efforts to dismiss the findings as methodological artifacts have failed. The gender differences in jealousy are robust across cultures, robust across methods, and not explained by alternative hypotheses. Given the current replication crisis in social psychology, these empirical success-stories are noteworthy. Importantly, none but the evolutionary psychological hypothesis led to these novel discoveries, a testament to the heuristic value of evolutionary psychology.[31]

Not everyone thinks that the matter is as cut and dried as Buss's review suggested. One of the issues is this: if Buss's ideas are correct, one would expect women and men—or heterosexual women and men, at least—to be sensitive to different characteristics in potential rivals. Women should be more jealous than men when their partners pay attention to physically attractive individuals, individuals with whom their partners would be most likely to desire sexual contact. Men should be more jealous than women when their partners pay attention to socially dominant individuals, those who can provide resources. A 1998 study by the Dutch psychologists Pieternel Dijkstra and Bram Buunk found exactly that.[32] In 2020, however, Thomas Pollet and Tamsin Saxton of Northumbria University in the United Kingdom published a study in which they largely failed to replicate the findings of the Dutch study. They did find a weak sex difference with regard to a rival's attractiveness but no sex difference at all with regard to a rival's dominance.[33] If nothing else, this shows that the so-called replication crisis in social psychology has not yet run its course.

Another issue has to do with sexual orientation and other characteristics that may bear on jealousy. According to several studies, including a very large 2020 study conducted in Brazil, nonheterosexual people do not show the same sex difference in jealousy as Buss reported for heterosexual men and women, largely because they are less concerned with sexual infidelity.[34] The same is true for people in consensually nonmonogamous (e.g., "open" and "poly") relationships. Other researchers have reported that the sex difference in jealousy varies with many factors, such as age, marital status, presence of children in the household, and experience of infidelity.[35]

Of course, evolved sex differences need to be flexible to meet different circumstances, but if they become too flexible, it is hard to distinguish them from the alternative, which is that they are learned. Although I'm a believer in the value of evolutionary psychology and think Buss's work exemplifies that value, much more research needs to be done to understand how inborn tendencies and social circumstances interact in the experience of jealousy for women and men.

CONCLUSIONS

Couples in established sexual relationships tend to resemble each other in a variety of demographic and personal characteristics. In part, this tendency toward homogamy may simply reflect the characteristics of the potential partners people are likely to encounter in the course of daily living. It's also possible, however, that people seek partners resembling themselves. Whether resemblance between partners benefits relationships over the long term is uncertain: some studies suggest that complementary relationships—for example, the combination of a more conscientious and a more extraverted partner—improve long-term relationship satisfaction.

Good communication is generally stressed as the primary basis for happy and durable sexual relationships. In this chapter, I discussed the work of one relationship expert, John Gottman, who has claimed the ability to predict the fate of couple's relationships based on analysis of their communication styles observed in brief laboratory settings. Based on this research, Gottman developed a therapeutic approach that he says can preserve happy relationships and rescue troubled ones. In my opinion, there are significant shortcomings to Gottman's research, and the benefits of his therapy programs, compared with other forms of relationship therapy, are not well established. Still, I focused on Gottman exactly because of his claims of scientific veracity; there's no reason to suppose that the more seat-of-the-pants approaches adopted by other therapists, counselors, and advice columnists get better results. Humans are complex as individuals, so the relationship between two humans is the product of two complexities, which makes for an astronomical enigma. Anyone who ventures to dissect—let alone repair—such an enigma has my admiration and, let's say, sympathy.

As with previous chapters, readers may wonder if I haven't forgotten about a central aspect of the topic—in this case, love, which by any accounting lies at the heart of happy relationships. I will present an extended discussion of love, but I postpone it to the final chapter because it's such an upbeat way to wrap up a book on sex. Before then I have to deal with four other topics. The first is that of paraphilias, those forms of sexual expression that lie on or beyond the fringes of conventional sexuality.

7

PARAPHILIAS

D o you know what chrematistophilia is? Not sure? Here's a clue: it's the same thing as harpaxophilia.

Still stumped? You're in good company. I didn't know what those words meant either, until a few minutes ago, when I came across them in *The Paraphilias*, a 2020 book written by Paul Fedoroff, a forensic psychiatrist at the Royal Ottawa Health Care Group in Canada.[1] Apparently both terms refer to sexual arousal from the experience of being robbed. There is a slight difference: harpaxophilia is sexual arousal by any form of robbery, whereas the term "chrematistophilia" may be used more narrowly to mean sexual arousal from being forced to pay for sexual services that you thought were being provided pro bono. That happened to me once, but I didn't find the experience sexually arousing; in fact, it put a damper on the entire proceedings. So I don't have chrematistophilia.

A PARAPHILIA is an unusual and persistent sexual desire or pattern of sexual behavior. The term has a medical flavor, and traditionally any person with a paraphilia was considered to have a psychiatric disorder. That changed in 2013 with the publication of the fifth edition of the American Psychiatric Association's *Diagnostic and Statistical Manual of Mental Disorders*. According to *DSM-5*, as it's called, a paraphilia is no longer to be considered a mental disorder unless it causes distress to the person who experiences it or harm to the persons affected by it, in which case it is called a PARAPHILIC DISORDER. So paraphilias range from healthy but unusual forms of sexuality, known commonly as "kinks," to a smaller number of sexual desires, such as pedophilia, that, if enacted, would likely incur severe criminal penalties. (I discuss pedophilia separately in the following chapter.)

The variety of paraphilias seems to be limitless. Fedoroff's book lists 149 of them; they include the familiar ones—shoe fetishism, masochism, voyeurism, and pedophilia—as well as many that must be uncommon, such as formicophilia (sexual arousal from being crawled on by ants), hodophilia (from traveling to new places), necrobestialism (from dead animals), and the above-mentioned chrematistophilia.

There are plenty of other paraphilias beyond those listed by Fedoroff. Another textbook, by Anil Aggrawal, lists 549 paraphilias.[2] Katharine Gates's 2017 book, *Deviant Desires*, mentions some rarities, such as sexual arousal from hiccups, balloons, and eye patches.[3] And I once received a letter from a man who told me that he was sexually excited by listening to the long words that I had used in a lecture. I was pleased to learn that I had done more than just educate people, but I didn't take him up on his request for a private hearing. In case this kink doesn't already have a formal name, I propose the term "polysyllabicophilia."

Homosexuality, at least in its garden-variety manifestations, is not considered a paraphilia even though it's hardly more common than some other sexual interests that are listed as such. The reason is that homosexuality was specifically delisted as a mental disorder by the American Psychiatric Association in 1973. Now that most paraphilias have themselves been delisted as mental disorders, however, there's really no reason why homosexuality shouldn't be readmitted to the category.

It's commonly stated that men are much more likely to experience paraphilias than women. Certainly the vast majority of people whose paraphilias bring them to the attention of law enforcement or mental health professionals are male. One paraphilic behavior that does engage women in considerable numbers, however, is the dominatrix role in BDSM (bondage, dominance, sadism, masochism, and other discipline-related activities). Women may take this role simply as a way to make money or out of curiosity, but it often becomes much more than that. One woman, Stavroula Toska, has described how she took on a job as a dominatrix for a few weeks with the intention of making a documentary film about BDSM, but she ended up continuing in the role for five years. "Here I was," she wrote, "a 34-year-old woman who was rediscovering herself and healing the wounds she never knew she had, all while making close to $2,000 a week."[4]

It's likely that more women have paraphilic interests than is generally believed. According to a 2022 study by Anna Madill and Yao Zhao of the University of Leeds in the UK, women's paraphilias are simply less visible because they are expressed mostly in fantasy and by the consumption of online videos rather than by overt behavior.[5] Among the paraphilic interests cited by the women in Madill and Zhao's survey were some that are fairly commonly reported by men, such as sex with minors, BDSM, and sex with animals, as well as some that were unusual or realizable only through fantasy, such as sex with ghosts and with pregnant men. If fantasy-only sexual interests qualify as paraphilias, then women are surely just as likely to be paraphilic as men. In one Canadian survey, for example, 65 percent of women reported having fantasies of being dominated sexually—an element of BDSM—whereas only 53 percent of men reported the same fantasy.[6]

WHAT CAUSES PARAPHILIAS?

When people who have a paraphilia are asked about its cause, they often attribute it to a particular early experience. Here's how one man recalled the onset, early in puberty, of a shoe FETISH: "I was home alone and saw my uncle's new penny loafers. I went over and started smelling the fresh new leather scent and kissing and licking them. It turned me on so much that I actually ejaculated my first load into my pants and have been turned on ever since."[7] Some sex researchers, such as Janet Shibley Hyde of the University of Wisconsin–Madison, have taken this and similar accounts as evidence for one-shot learning as a cause of paraphilias.[8] Orgasm is certainly a rewarding experience, and as such it could be the positive reinforcement in a conditioning paradigm. Still, this writer's account makes it seem like he already had a shoe fetish, or a predisposition to develop one, before he even saw the loafers. Otherwise, why would he have gone over to them, smelled them, kissed them, licked them, and been so sexually aroused by them that he ejaculated without even touching himself? And if paraphilias do develop by one-shot learning, why doesn't everyone become sexually fixated on whatever person, event, or sensory experience is associated with their first orgasm?

Quite a few attempts have been made to create paraphilias by CONDITIONING in the laboratory, and several have shown positive effects. The most-cited study is by the British psychologist Stanley Rachman published in 1966.[9] He repeatedly showed three male psychologists a photograph of knee-length women's boots followed by a photograph of an "attractive, naked girl"—a young woman, presumably—while he measured the diameter of their penises with a device called a "penile plethysmograph." Initially the boots triggered no arousal, but after repeated pairing with the erotic photo, they began to do so, even when presented without the erotic photo. This could be considered classical conditioning. It's analogous to Ivan Pavlov's famous experiments on dogs in which he repeatedly paired the ringing of a bell with the presentation of food: the dogs initially salivated only in response to the food but eventually came to salivate at the sound of the bell presented by itself.

None of the published studies has reported the creation of a fetish with anything like the strength or durability of the real thing. But the volunteers have always been adults, who presumably had plenty of prior sexual experience. It could be that there's a window of susceptibility to paraphilias coinciding with puberty or a person's early sexual experiences. If so, to demonstrate a strong conditioning effect, it would be necessary to carry out these experiments on sexually naive boys or girls. Such experiments are ruled out by ethical considerations, of course.

One of the best-known proponents of the conditioning theory of paraphilias is the behavioral neuroscientist James Pfaus, who is currently associated with Charles University in Prague. Pfaus and his colleagues studied the effects of allowing male rats to copulate with females that had been imbued with some unnatural characteristic, such as an almond scent. When later given the choice of copulating with an almond-scented or an unscented female, the male rats chose the scented female. This was true only for sexually naive males, however; allowing a male prior sexual experience with unscented females prevented the development of the "fetish."[10]

The findings of animal experiments like this one are somewhat consistent with reports, such as the one cited earlier, in which paraphilias are recalled as having arisen in conjunction with a person's initial or early sexual experiences. Still, a human with a strong shoe fetish will be highly aroused by a shoe without an attached person, and they will not be highly aroused by a person without an attached shoe; neither of these conditions was replicated in Pfaus's experiments.

A person with a single paraphilia may develop others over the course of their lifetime. According to one study, the majority of people with paraphilias, by the time they come to the attention of clinicians, report having developed more than one, and 18 percent have developed four or more.[11] But the fact that these additional paraphilias appeared over the course of a person's adult life doesn't mesh well with the idea of an early "susceptibility window." Rather, the accumulation of paraphilias in particular individuals suggests the existence of some predisposing factor within those individuals, a factor that continues to operate—perhaps with increasing force—over their lifetime.

Genes predisposing to paraphilias could be such a factor. Paraphilic disorders do appear to run in families, and this clustering is partly genetic in origin, according to family and twin studies.[12] Still, the genetic influence is weaker for paraphilic disorders than for many other mental disorders or nonpathological psychological traits, and specific genes have not been identified. There is little evidence regarding a possible genetic influence on nonpathological paraphilias (kinks).

Another possible causal factor could be a lack of social skills. Some paraphilic disorders, such as voyeurism, exhibitionism, and frotteurism (groping), seem designed to bypass the need for ordinary social interactions. For that reason, social skills training is often included in treatment programs for these disorders.

It's also possible that some paraphilias are adopted because they allow a person to hide aspects of their true selves that they're not comfortable with. That could be true for men who enclose themselves in multiple layers of rubber, for example, or for furries—men or women who dress in "disneyfied" animal costumes (fursuits). Only about one-half of male furries say their motivation is primarily sexual, however.[13] Debra Soh, then a graduate student at York University in Toronto, attended a furry convention, interviewed many furries, and summarized her findings as follows "The one message that was consistent across my conversations was that each member of the community felt they had something that made them different and ill-fitting in mainstream society, such as Asperger's syndrome or a facial tic. They found some aspect of childhood, such as cartoon characters or stuffed animals, to be comforting, and this appreciation continued on into their adult lives. The fandom gave them a safe venue in which to express themselves and to feel accepted by others who feel similarly."[14]

There are occasional reports of paraphilias developing following brain injuries. In his book, Fedoroff cites the case of a man who, after a brain injury sustained in a motorcycle accident, could experience orgasm only while wearing nylon stockings. This is how the man described waking up in the hospital: "I was lying on my back with a brace on my neck so I could hardly move. Then a nurse walked into the room and I could hear her nylons rubbing together. I have never been so turned on and I could not even see the nurse. It was the sound of her nylons. Since then, nylons are all I think about when I think about sex. . . . Before, if I knew a guy liked nylons I would beat him up. Now I am one of them!"

BECOMING THE BELOVED

There's a curious phenomenon experienced by some men with paraphilias and, occasionally, men with more conventional sexualities. That is a tendency to incorporate features of people to whom one is sexually attracted into one's own self-image. In other words, a man may be sexually aroused by imagining himself as, or actually turning himself into, the kind of person he finds sexually desirable.

I just mentioned the man who, following a brain injury, could experience orgasm only while wearing nylon stockings. It's common for straight men with clothing fetishes to put on the clothing that arouses them, even though the clothing is likely to be female-typical attire. You could think of it as a kind of short cut, one that eliminates the need to identify and establish sexual contact with women clothed in the desired fashion. The great majority of such men have no history of brain injury.

Some heterosexual men are sexually aroused by wearing complete female attire, and they may masturbate while so dressed. This phenomenon is known as "heterosexual transvestic fetishism," or HETEROSEXUAL TRANSVESTISM. It's different from the phenomenon of individuals who are born anatomically male and whose lifelong gender identity is female, because in the latter case, wearing female attire is an expression of gender identity and not in itself a source of erotic arousal. Men with heterosexual transvestism may wish to have sex with female partners—their wives, for example—while wearing items of feminine attire. Women are frequently turned off by such requests, however, and may wonder if their partners are actually gay.

Here is an account of one heterosexual transvestic fetishist, "Mario," excerpted from a clinical study by Kenneth Zucker and Ray Blanchard:[15]

> At the age of 11 or 12, Mario began dressing in his sister's clothes when no one was home. He initially wore lingerie, but later came to include dresses and makeup on occasion. Mario was sexually naive at this point and did not understand the arousal he felt when he put on women's clothes.

Mario's first ejaculation occurred at age 14. He was lying face down on his bed wearing a brassiere and panty hose and examining the lingerie pages of a department store catalog. While studying a photograph of a model with panty hose like those he had on, he began unconsciously to thrust against his mattress, with resulting ejaculation. In later life, he continued to find women in lingerie more attractive than nude women and to be more aroused by lingerie advertisements than by pornography. The young Mario realized that there was something unusual about his sexual behavior and wondered for a time if he were homosexual.

In later adolescence and adulthood, Mario's transvestic and masturbatory activities were accompanied by fantasies of sexual interaction with women. In one favorite fantasy, Mario would cross-dress with a woman's permission (sometimes at her insistence) and then have sex with her in a quasi-lesbian interaction. Mario was also sexually aroused by the sight of himself cross-dressed in the mirror.

Although there is some similarity between this account and the penny loafer story mentioned earlier, I find this one to be more believable, especially because Mario's fetishistic interest preceded his first orgasm rather than being the consequence of that event.

In some men, heterosexual transvestism proceeds further, to the point that the men are aroused by simulating attributes of female anatomy; they may be aroused by wearing false breasts or false vaginas, for example. Finally, a few men seek breast implants, genital surgery, hormone treatments, and other procedures with the aim of actually becoming biological women, or as close to becoming women as medical technology permits.

Although that is a form of transexuality, it's different from the better-known "classical" transexuality, in which the motivation for medical assistance with transitioning is affirmation of one's internal gender identity rather than sexual arousal. (For that reason, the term "gender affirmation surgery" is generally preferred to "sex change surgery.") Much of the work on this form of transexuality has been conducted by Ray Blanchard, who gave it the name AUTOGYNEPHILIA, meaning sexual arousal in a biological male from imagining or recreating oneself as a woman.[16]

The sexual orientations of autogynephilic and classical trans women are usually different. Most autogynephilic trans women are sexually attracted to women, whereas most classical trans women are sexually attracted to men. Another difference between autogynephilic and classical transexuality is developmental: during their childhood, most classical trans women identified as girls or expressed the desire to become girls, whereas most autogynephilic trans women identified as boys, in conformity with their anatomy at birth, and did not seek to transition until later in life after going through the stages of "internalization" just described.

Some trans activists have objected to the concept of autogynephilia, one frequent complaint being that it pathologizes transexuality by representing it as a paraphilia.[17] In my opinion, it's reasonable to call autogynephilia a paraphilia, but only if one accepts the view, codified by *DSM-5*, that paraphilias are not disorders unless they are specifically designated as such. Of course, the Greek-sounding term autogynephilia does give the concept a clinical flavor, whether or not it is officially a disorder.

There are also plenty of trans women who believe that the concept of autogynephilia is an appropriate description of their own identity and sexual history.[18] It would be helpful if autogynephilic individuals were to adopt a more user-friendly moniker for their identity, just as men-loving men prefer the term "gay" to the more clinical "homosexual."

One can find many other examples of this kind of internalization of the object of one's erotic desires. Some were described and analyzed in 2009 by the sexologist Anne Lawrence, who at the time was associated with the University of Lethbridge in Alberta.[19] There's a paraphilia called "acrotomophilia," for example, in which sexual attraction is directed toward persons with amputation stumps. In some individuals, that attraction is internalized, so the person is aroused by the thought of having an amputation themselves ("apotemnophilia") or by mimicking an amputation, and occasionally such a person may seek to have a limb removed surgically. (In some cases, the desire for the removal of a healthy limb may be driven not by sexual arousal but by a defect in the representation of that limb in the brain, such that the limb is not accepted as belonging to the self.)

Similarly, some men who are sexually attracted to prepubescent or pubescent children may be sexually aroused by mimicking physical and psychological characteristics of children to whom they are attracted. The sexologist Michael Bailey of Northwestern University, whose research into sexual orientation was discussed in chapter 4, has proposed that pop star Michael Jackson was such a person. Jackson's facial surgery, his adoption of a high-pitched voice, and other aspects of his behavior were certainly suggestive of such an interpretation, as was his statement, "I am Peter Pan."[20]

Do some gay men also internalize the objects of their desire? That's a tricky question to answer, because gay men already belong to the sex that they find attractive, so they don't have to imagine or do anything to resemble their sexual "targets." And if, as many gay men do, they work out to make their bodies more attractive, "more attractive" might mean more attractive to other men or more attractive to themselves, or both. A gay character in a novel by Alexander Eberhart captures this ambiguity: he recalls his earliest attraction to a well-built friend of his older brother, musing, "Did I want to be *like* Phillip, or did I want to be *with* Phillip?"[21]

Gay men may indeed find their own bodies sexually arousing. Here's one snippet of evidence. In a survey of German college students, 48 percent of gay male students said that they had masturbated in front of a mirror within the previous six months, whereas only 17 percent of straight male students said the same thing.[22] Yet something prevents most gay men from becoming the exclusive focus of their own desire. That something could be the Westermarck effect, described in chapter 6. It also could be the effect of habituation, because novelty is an important factor in stimulating and maintaining sexual attraction and arousal.[23]

The phenomenon whereby the target of one's sexual attraction is incorporated into oneself has been given technical names: EROTIC TARGET LOCATION ERROR, or "erotic target identity inversion." Both "error" and "inversion" are loaded terms. "Error" states outright that something is wrong, and distinguishing between "inverted" and "right way up" is largely subjective: if Australians had invented cartography, for example, the South Pole would be at the top of the map. The only merit to these terms is that they make clear we are dealing with a general phenomenon, something that applies to a minority of individuals within many—perhaps all—forms of sexual expression. In that sense, the direction of sexual attraction, to an external target or a representation of that target located within oneself, can be thought of as a dimension of sexuality, one that is independent of other dimensions such as sexual orientation.

CONCLUSIONS

It's difficult to make any generalizations about paraphilias, because they encompass such a broad range of desires and behaviors, some of which are harmless or even creative and enriching, whereas others are distressing, traumatizing to victims, or heavily stigmatized. Correspondingly, the people who experience or enact paraphilias are a highly diverse bunch, ranging from sexual adventurers to socially impaired misfits and deeply disturbed individuals.

The boundaries between healthy and pathological paraphilias are unclear and may shift over time. For example, the internet has helped people with unusual sexual interests join like-minded communities; in doing so, their distress may evaporate, thus eliminating a criterion for their diagnosis as having a mental disorder. Some unusual sexual interests, such as those that fall under the catch-all abbreviation BDSM, have gone mainstream, even becoming fashion trends, perhaps to the vexation of hardcore practitioners.

Even one of the most secure generalizations about paraphilias—that they are more common among men than women—is thrown into doubt if, as just discussed, paraphilias are experienced more privately in women than in men—in fantasies rather than in observable behavior.

At least we can be sure that all paraphilias are sexually motivated, because that's part of the definition. Or is it? If some of the attendees at a furries convention are looking for sex and some are not, yet all are dressed and acting the same way and socializing enthusiastically with one another, then that raises a question as to whether furry fandom is really about sex at all. Wouldn't that be like saying that having sex under canvas is a paraphilia because some hikers like to have sex on their overnight trips?

Even quite a few BDSM practitioners don't engage in genital contacts or any other behaviors that would be generally recognized as sexual; for them, BDSM may be about power or control rather than about sex. Even spirituality is, for some, the central meaning of BDSM.[24]

The fact that, as reported, individuals with one paraphilia are likely to develop several others over the course of their lives suggests that, for some people at least, there is an underlying tendency to sexualize experiences or interests that are not inherently sexual and to focus their energies on these interests in an obsessive, autistic, or addictive fashion. I'm certainly not qualified to evaluate the relevance of such diagnoses to paraphilias. (I do make some comments on sex addiction in chapter 9.) The point, though, is that the diversity of paraphilias—whether there are 149 of them (per Fedoroff), 549 of them (per Aggrawal), or some even larger number—may be irrelevant and distracting if the core feature of paraphilia is a personality trait that is prone to sexualize anything, sexual or nonsexual.

One paraphilia does stand out from the rest, due to the overwhelming concern it arouses in the public sphere. That paraphilia—sexual attraction to children—is the topic of the next chapter.

8

PEDOPHILIA

Whereas kinks and even some paraphilic disorders such as exhibitionism and voyeurism are frequently the targets of curiosity or humor, there is nothing light-hearted about PEDOPHILIA. This term refers to sexual attraction that is directed toward children who have not yet reached puberty or, in a somewhat broader definition, toward children no older than thirteen. (Most thirteen-year-olds have already entered puberty.)

Not all pedophiles have sexual contact with children, and not all adults who have sexual contacts with children are pedophiles. Still, pedophilia is a serious problem: if enacted in behavior, it risks harming its victims and landing its perpetrators in prison. If not enacted, it is likely to be a lifelong psychological burden.

The great majority of known pedophiles are men, but female pedophiles do exist, and they have to contend with being a more or less invisible minority within a larger, highly stigmatized minority—a "double freak," as one women put it to interviewers.[1] The number of female pedophiles would be higher if we were to include women who report having fantasies involving sex with children, as mentioned in the previous chapter.

WHAT CAUSES PEDOPHILIA?

As with paraphilias in general, numerous ideas have been presented about the cause or causes of pedophilia. One very prevalent idea is the so-called sexual-abused-to-sexual-abuser hypothesis, or simply the abused–abuser hypothesis. According to this hypothesis, a child who experiences sexual contact with adults may later desire to have sexual contact with children—to repeat the abuse, in other words, but with the roles reversed. It's not obvious why a person who's had a bad experience should want to inflict that same bad experience on others, but several explanations have been offered for why that may be so, such as a supposed tendency to identify with one's abuser.

Sexual offenders against children (who may include pedo-philes and nonpedophiles) are more likely to report having been sexually victimized as children than do nonsexual offenders, according to several studies. These observations have been taken as evidence in favor of the abused–abuser hypothesis. It's pos-sible, however, that child molesters may remember episodes of sexual abuse better than other offenders, or they may be more likely to declare them, perhaps as a result of having been specifi-cally questioned on the topic. They may even invent memories of sexual abuse as a way to explain or excuse their crimes. Further-more, the abused–abuser hypothesis runs up against an incon-venient truth: most sexually abused children are girls, whereas a very large majority of known pedophiles are men.

To circumvent the problem of biased recall, a group at Griffith University in Queensland, Australia, led by the forensic psychologist Chelsea Leach, carried out a longitudinal study of all boys born in Queensland in the years 1983 and 1984.[2] The data came from child protection reports and adolescent and adult court records. There were 286 boys who experienced sexual abuse but no other form of abuse; they were mostly around the ages of seven to nine when the abuse first occurred. Of these boys, only two committed sexual offenses of any kind by the age of twenty-five. Another 329 boys experienced sexual abuse in combination with physical or emotional abuse or neglect; of these, only seventeen committed sexual offenses. Conversely, of all the boys in the sample who eventually did commit a sexual offense, only 4 percent were reported as having experienced any sexual abuse during childhood. Not surprisingly, then, statistical analysis failed to reveal any specific association between childhood sexual victimization and later sexual offending.

A similar longitudinal study, carried out in Australia, also found that only a tiny minority—1 percent—of sexually abused children went on to commit any kind of sexual offense.[3] Despite the low numbers, the sexually abused boys were somewhat more likely to commit sexual offenses (not necessarily against children) compared with nonabused boys. It was not determined, however, whether these boys experienced other forms of abuse in addition to sexual abuse.

All in all, the evidence in support of the specific sexual-abused-to-sexual-abuser hypothesis is weak. It is quite possible, however, that boys subjected to multiple forms of abuse that may include sexual abuse are at increased risk of committing a variety of crimes, including sex crimes, in adulthood. Considering the deplorable social conditions in which multiply-abused boys are likely to have been raised, you'd think that they would inevitably slide into a life of criminality. But that outcome is far from inevitable. One counterexample is offered by the former U.S. senator for Massachusetts, Scott Brown. In his autobiography, Brown described a childhood in which he experienced repeated sexual and physical abuse and much more; yet after an adolescence marked by some antisocial behavior, he became an outstanding public servant. "Like a fractured bone, I have knit back stronger in the broken places," Brown wrote.[4]

Another idea is that pedophilia is caused by some quirk of brain structure or function. One well-known exponent of this idea is James Cantor, a sexologist at the Centre for Addiction and Mental Health in Toronto. Cantor believes that events during prenatal life, starting as early as five weeks of gestation, mark a developmental pathway leading to pedophilia.[5] He bases this conclusion on reports from his research group that possible markers of atypical prenatal development— non-right-handedness, minor craniofacial anomalies, short stature, and low intelligence—are more common among pedophiles than among men attracted to adults. Within the brain, Cantor's group reported finding atypical organization of the white matter—the axonal tracts that interconnect brain regions—in the left hemisphere of pedophiles compared with nonpedophilic men.[6] Some studies from other groups have found atypical features of the grey matter in pedophiles, but those findings have not been consistent from study to study.

Abundant caution needs to be shown in the evaluation of Cantor's ideas. For example, most of his studies have been based on men charged with or convicted of sexual offenses against children. One could imagine that other pedophiles—those who have not offended or who have offended but evaded detection— might differ in significant ways from the convicted offenders.

To explore this idea, a German research group compared the ability to inhibit behavior in offending and nonoffending pedophiles using psychological tests and functional brain mapping. They found, as one might expect, that the nonoffending pedophiles had superior behavioral control: they were better able to avoid responding to the incorrect images in a rapidly presented sequence of images.[7] Another recent study, also from Germany, reported that most of the characteristics linked by Cantor to pedophilia were present in offending but not in nonoffending pedophiles, and those authors suggested that Cantor's characteristics are factors that predispose to criminal behavior in general, not to a sexual interest in children.[8]

All in all, the findings of Cantor and other groups suggest, but do not yet nail down, a biological basis for pedophilia. Part of the reason for the lack of certainty is simply the dearth of information: far less biologically focused research has been done into pedophilia than into the sexual orientation of adults. And none of the present findings provide a means to tell whether a person who denies being a pedophile is telling the truth. As Cantor himself wrote, "The statistical relations of IQ and handedness to pedophilia, although valuable as potential clues to the etiology of this disorder, are far too small to permit these variables to be used as diagnostic indicators."

A number of other possible causes of pedophilia have been proposed, including the ideas mentioned in the previous chapter in connection with paraphilias in general: conditioning, an impaired ability to interact socially with adults, genes, and brain injuries. The researchers who developed a maternal immunity theory for male homosexuality have proposed a similar mechanism to explain pedophilia directed toward boys.[9] A limited amount of evidence has been put forward in support of all these ideas, and they remain to be explored in depth. In a word, we don't yet know what causes pedophilia.

IS PEDOPHILIA A SEXUAL ORIENTATION?

As I discussed in chapter 4, the term "sexual orientation" traditionally refers to the balance of sexual attraction to males and

females. Like people who are sexually attracted to adults, pedophiles have sexual orientations; typically they are attracted more
strongly to either female or male children. But many pedophiles,
as well as some sex researchers and psychiatrists, believe that
pedophilia should be considered a sexual orientation in itself,
equivalent in some ways to homosexuality and heterosexuality,
rather than as a paraphilic disorder.

In one survey of adults who are sexually attracted to children, conducted by the sociologist Allyn Walker, who was then
at Old Dominion University, most participants said this attraction was their sexual orientation or sexuality.[10] They did not use
these terms as a way to justify sex with minors, however. As one
participant told Walker, "I think it's an orientation, but not an
orientation that can ever be acted on. But I think it's an orientation." (As a result of this and related research, Walker was forced
to resign from Old Dominion in 2021 but was hired by the
Moore Center for Prevention of Child Sexual Abuse at Johns
Hopkins University.[11])

The term "sexual orientation" can be qualified to cover other
aspects of sexual attraction, as, for example, with the phrase "age
sexual orientation." That phrase was introduced—or at least
popularized—by Michael Seto, a forensic psychologist at the
Royal Ottawa Health Care Group. According to Seto, there is
a spectrum of age orientations, which he calls CHRONOPHILIAS,
whose targets range between the extremes of infants at one end
to very old people at the other.[12] There is a range of technical
terms referring to waystations along this spectrum: "nepiophilia"
(attraction to infants), pedophilia (prepubescent children),
"hebephilia" (pubescent children), "ephebophilia" (postpubescent
minors), "teleiophilia" (young or middle-aged adults), "meso-
philia" (specifically middle-aged adults), and "gerontophilia" (old
people). The great bulk of people are, of course, attracted most

strongly to younger or middle-aged adults and are therefore teleiophiles; only very small numbers of people are nepiophiles or gerontophiles. The boundaries between the age-based categories are not sharp: many or most pedophiles experience some attraction to pubescent children, for example.

Seto points out that there are parallels between age sexual orientation and the traditional sexual orientation toward males or females, which for clarity we can "gender sexual orientation." Looking specifically at pedophilia, Seto mentions several factors. For one thing, most pedophiles become aware of the targets of their sexual attraction at the same age as do other people, namely, around the time of puberty, and—according to most experts—this attraction remains stable over the person's lifetime and does not readily change even if the pedophile strongly desires to change.

Second, according to an online survey of three hundred pedophiles and hebephiles conducted by Seto and his colleagues, it's typical for these individuals to experience both sexual and romantic attraction to children, just as teleiophiles experience both forms of attraction to adults. In fact, over 70 percent of the individuals in the survey stated that they had fallen in love with a child.[13] In addition, Seto points out that the patterns of brain activation seen when pedophiles view potentially arousing images are similar to the patterns seen in teleiophiles when they view arousing images of adults of their preferred sex.

Seto does not think that labeling pedophilia an orientation legitimizes sex between adults and children, any more than most pedophiles do, as the survey mentioned earlier revealed. He does, however, believe that pedophiles should be respected rather than reviled for who they are, and he has a particularly positive view of the group Virtuous Pedophiles, which consists of, supports, and advocates for nonoffending pedophiles.[14] ("Virtuous," in this context, means committed to avoiding sexual contacts with children; the group doesn't believe that pedophiles are more virtuous than anyone else.)

The website for Virtuous Pedophiles includes numerous statements by people seeking membership in the group. They almost uniformly speak of the pain they have experienced; thoughts of suicide are frequently mentioned. Yet it is likely that some nonoffending pedophiles—especially those who have received support from Virtuous Pedophiles or other organizations—no longer experience distress and are reconciled to their situation. Such individuals, who are not distressed and do not offend, should not be considered to have a mental disorder at all, according to Ray Blanchard, who cowrote the section of *DSM-5* dealing with paraphilias.[15] Not all sex researchers agree with him.

Paul Fedoroff, author of *The Paraphilias*, mentioned in chapter 7, is at the same institution as Michael Seto, but their views on pedophilia could hardly be more different. For starters, Fedoroff rejects the idea that pedophilia is a sexual orientation. To him, sexual orientation means the direction of a person's romantic attractions. Pedophiles have no romantic attraction to children, according to Fedoroff, and therefore that's not their orientation. They simply view children as desirable sex objects with whom they want to have physical sexual contact. Fedoroff calls that a sexual *interest*.

When Fedoroff explained these ideas to me in a 2021 interview, I was taken aback. The usual definition of sexual orientation—the direction of a person's sexual attraction to males, females, or both—doesn't specify either romantic or physical attraction, but the two usually go together. If 70 percent of Seto's pedophile interviewees have fallen in love with a child, then the two aspects of attraction seem to go together for many or most pedophiles, just as they do for most other men and women.

I mentioned two names to Fedoroff: Charles Dodgson and Michael Jackson. Dodgson, better known by his pen name Lewis Carroll, wrote the Victorian children's classic *Alice's Adventures in Wonderland*. Dodgson's devotion to the real-life Alice, and to other young girls, is well documented, as is Jackson's devotion to a series of young boys. That these men's devotion included sexual desire is unproven but a matter of strong surmise. When I brought up these two names, Fedoroff conceded that some pedophiles' attraction to children might be romantic as well as physical in nature, but these would be a small minority of all pedophiles, he said.

CAN PEDOPHILIA BE CHANGED?

Fedoroff shares the widespread belief that a person's sexual orientation is difficult or impossible to change. But that's irrelevant to pedophilia, Fedoroff told me, because pedophilia is not a sexual orientation but a sexual interest. And simply explaining this fact to pedophiles—that they do not have to change their orientation—is an important first step toward refocusing their sexual interest on to adults.

I would dismiss this idea out of hand were it not for one thing: Fedoroff has seen and treated many pedophiles over the course of a long career. And he claims to have had great success in helping those people swap their sexual interest in children for an equivalent interest in adults—usually adults of the same sex as the children on whom they were originally fixated.

Fedoroff backs up this assertion with data he obtained in a 2014 study led by Karolina Müller, who was then a master's student.[16] Fedoroff's group focused on forty-three men who had been diagnosed with pedophilia, most of whom had committed sexual offenses against children. The men's physiological arousal patterns were measured on at least two occasions: at the time of initial diagnosis and six months or more later, after a period of treatment. Initially nearly all the men showed arousal—that is, some degree of penile erection—while listening to accounts of erotic interactions with children but not to comparable accounts of interactions with adults. That being so, these men were likely to have been pedophiles and not men who sexually molested children for other reasons.

At the time of the later session, about half the men showed the reverse pattern of arousal, responding more strongly to accounts involving adults than to those involving children. In other words, there was objective evidence that their sexual interests had changed to more typical and acceptable targets. Those men apparently were no longer pedophiles. Even the men who failed the treatment, in the sense that they continued to show sexual arousal to children, became less likely to reoffend. In Fedoroff's clinic, "the incidence of hands-on sexual reoffenses against children has been virtually unknown during the past 15 years," Fedoroff wrote in *The Paraphilias*. (He meant, of course, that reoffending was unknown, not its incidence.)

What was the treatment that produced these remarkable changes? In individual psychotherapy, Fedoroff first made it clear that the men must immediately cease all illegal acts. He then helped the men abandon false beliefs about sexual contacts with children, most particularly the idea that children welcomed and benefited from such contacts. The men also participated in group psychotherapy sessions. Those groups comprised a mix of recently diagnosed pedophiles and men who had already experienced some degree of change; thus the "newbies" were able to learn directly from the "old hands" that change is possible.

I have difficulty accepting these results at face value. That's partly because back when homosexuality was listed as a mental disorder, there were many claims of techniques to cure it— techniques that are now widely dismissed as bogus. More specifically, some experts have critiqued Fedoroff's study on a variety of theoretical or technical grounds, such as his supposed application of inappropriate statistical procedures.[17] A 2020 review of relevant studies, by Ian McPhail and Mark Olver of the University of Saskatchewan, concluded that in men convicted of sexual offenses against children, therapy can help damp down their sexual arousal to children but cannot create arousal to adults.[18] Fedoroff has published detailed rebuttals to all these critiques. His critics are so wedded to traditional ideas about pedophilia, Fedoroff believes, that their minds are closed to any novel approach.[19]

I asked Fedoroff what he thought of the support group Virtuous Pedophiles. Most people working in the field believe that the group's work is of value both to pedophiles and the wider society. But Fedoroff told me he feels sorry for pedophiles who join the group. By buying into the false belief that pedophilia is an immutable orientation, Fedoroff said, these men are cutting themselves off from the real possibility of recovery and the reestablishment of a normal and healthy sexuality.

A PEDOPHILE'S VIEWPOINT

Todd Nickerson is a forty-nine-year-old graphic artist who lives in rural Tennessee. He is a member of Virtuous Pedophiles, where he moderates online discussions. He identifies as a nonoffending pedophile and also as a "MAP." That's an acronym for MINOR-ATTRACTED PERSON—an umbrella term that includes pedophiles as well as adults who are attracted to older children or adolescents. Nickerson "came out" as a pedophile in a 2015 article published by *Salon*. The article was later deleted from the magazine's website, but it is still available on an archival site.[20]

In a 2021 Skype interview, I asked Nickerson about his sexuality. "My primary attraction is to prepubescent girls," he said,

but occasionally I'm attracted to boys as well. Kids are pretty androgynous. I'm attracted to what I perceive to be feminine qualities in children, which can be seen in some boys—softness. I'm not really picky about genitalia because I'm never going to experience that in any way.

For me, it's more that I'm attracted to personalities. It's a blend of sexual and romantic attraction. I consider myself to have fallen in love with two girls over the course of my life. I don't know why people think that if pedophiles like spending time with kids, it's just because they're grooming them for sex. It's not true, not for me anyway. It's almost like, emotionally, I regress to their level.

The first time I ever became aware that I'm not like everybody else—I was 11 or 12, and I was with some other boys, and one boy asked which girls we liked in the class. They all agreed that this one girl who was the most developed girl was the most attractive, and I kept my mouth shut because the girl I actually liked best was the least developed.

I did some babysitting when I was a teenager—and that's when I fell in love with a girl for the first time. It was when I was 18. I had to leave that job because it was messing with my head too much. Up to that point I was kind of in denial.

In my 20s I was trying to force myself to be attracted to adults, but it didn't work. Right after college I spiraled into a long-term depression. I was thinking about suicide on a daily basis. I went online, but there wasn't a lot of information. I ended up on this forum, girlchat. Most of the people in that community were pro-contact [i.e., they favored the legalization of sex between adults and children]. When you're extremely depressed it's easy to get sucked into that, it becomes very cultlike. I started espousing their talking points. It lasted about a year before I woke up and thought, "This isn't right." I started arguing with them, and they turned on me.

I asked Nickerson what he thought had caused him to become a pedophile. "I feel like it's some kind of genetic condition that lies dormant, but in some people it gets activated by certain experiences. I did have one sexual contact with an adult when I was 7. It just involved him touching my genitals. I didn't enjoy it, but I wasn't frightened or anything. I just thought it was odd. I didn't affect me much at the time. Later on, when I realized that this was a shameful thing, that's when I started having issues with it.

"People think that all pedophiles were molested when they were children, but I know quite a few who were never molested. A lot of them had sexual experiences with other kids—that's probably more common. MAPs who were molested, many of them know how harmful it can be and that's why they don't offend." Nickerson also mentioned a physical disability—he was born missing his right arm—which he believes contributed to the social isolation he experienced as a child.

I asked Nickerson whether he thought pedophiles' sexual attraction to children could be redirected toward adults, as claimed by Fedoroff. "I don't buy it. It's not a thing to be cured," he said. "It's not a disease, it's a sexuality. I've been to a lot of different counselors and unanimously they've said, 'I'm not going to try and change your sexuality, I don't think I can do it anyway.' Fedoroff is in a very small minority. The people who 'changed' were probably never pedophiles to begin with. Or maybe they had some pedophilic attraction but they were still attracted enough to adults. There are a lot of MAPs who are attracted to adults as well as to children."

Can pedophiles channel their attraction to children in nonsexual ways that are beneficial to children, such as through teaching? "It depends on the MAP," Nickerson said. "If they know they're not going to offend, I don't think I'd tell them 'Don't take a job as a teacher.' MAPs can be very good teachers, because they care. I know MAPs who are teachers who would never dream of offending, and they are very loved and very respected. But there's always that pesky thing of the ones who end up offending. Personally, I wouldn't do it.

"MAPS are a cross-section of society, just like anyone else— most of us are moral people, who don't want to cause any harm. They're stuck with this sexuality—I've known a couple who have taken their own lives. That's pretty common, especially among the younger MAPs. They don't have the 'It Gets Better' stuff like the LGBT community has now. Some call LGBT support lines, but they say: 'I can't support you, I'm not qualified on that issue.'

"Part of the purpose of Virtuous Pedophiles is to help people who have no support otherwise. If they're in a relationship they can't tell their partners, they can't tell their family. They're even afraid to tell their counselor—if they have a counselor—because of mandatory reporting laws, even if they haven't done anything. Pedophiles need to be able to talk freely about their situation. As I wrote in the *Salon* article, 'A repressed, unhappy pedophile is a pedophile at risk.'"

CONCLUSIONS

Pedophilia—sexual attraction to prepubescent or early-pubescent children—is quite a mystery. It's not clear how a person comes to be a pedophile: whether it's a consequence of sexual abuse during childhood, an atypical neurohormonal development, genes, or some other cause. Even more puzzling is how to think about pedophilia in evolutionary terms.

What's clear is that pedophiles are regular human beings whose sexual development parallels that of nonpedophiles, except for the atypical object of their desire. For that reason, calling pedophilia a sexual orientation, or chronophilia, seems reasonable to me.

I'm skeptical that pedophiles can be "cured"—converted to sexual attraction to adults—whether by Fedoroff's therapy program or by any other means. Nevertheless, it's important to keep an open mind: Fedoroff, or others working in the field, may eventually provide convincing evidence to support their claims.

I can't think of any reason to vilify or persecute those pedophiles or MAPs who have committed themselves not to offend sexually against children. On the contrary, they deserve admiration and support in maintaining what, for the rest of us, may seem like an unimaginably difficult life choice. And academics who express opinions such as these should not be hounded out of their institutions.

9

PORN

Pornography is huge. Just one website, Pornhub, welcomes 42 billion visitors annually. Every second, it delivers over two hundred gigabytes of sex-drenched videos. Just to watch twelve months' worth of newly added Pornhub content would take 169 years of uninterrupted viewing.[1] And every year, thousands of articles and books are added to the academic corpus of pornology.

I've watched an infinitesimal fraction of those videos, and I've read just a few dozen of those academic offerings. But I'm as well qualified to write about the topic as anyone else, because we are all blindly prodding at the elephant that is porn.

Porn is art; it's smut. Porn is healthy; it's a cancer. Porn is legal; it's a crime. Porn is fun; it's a drug. It's a blessing; a sin. Porn is whatever you see in it. It's lovely; it destroys love.

Porn is more than videos, of course. It's any material that is meant to be sexually arousing. Porn is photographs and paintings, sculpture and songs and stories, medieval woodcuts, graffiti in the latrines of Pompeii, pre-Columbian ceramics, maybe even prehistoric cave scribblings. But right now, Pornhub and its rivals so dominate the genre that streaming videos are almost everything that people understand by the term.

PORN'S BENEFITS

What's good about porn? One person who knows something about that is Taylor Kohut, a research associate at the University of Western Ontario. He has focused on porn for the last twelve years. In one study, he and two colleagues recruited more than four hundred men and women who were in heterosexual relationships—more specifically, relationships in which at least one partner used pornography.[2] Here are some of the benefits of watching porn, according to participants in the study.

- *Porn improves mood*: "Seeing this porn gives me a feeling of life and of joy—a zest for living."
- *Porn increases arousal*: "[It] helps keep our sex life exciting, which makes both of us happier."
- *Porn improves communication*: "It helps us facilitate a conversation about kinks, fetishes, wants and won'ts in the bedroom."
- *Porn provides information*: "I feel like pornography helped educate me on how to perform oral sex better."
- *Porn promotes experimentation*: "She enjoys using toys and positions that we were first exposed to by pornography."
- *Porn provides an alternative outlet*: "It can get her an orgasm when I am not available or not in the mood."
- *Porn reduces guilt*: "I believe that it has taught me that the body and sex are meant to be enjoyed."

Some of the participants did list negative consequences of watching porn:

- *Porn raises unrealistic expectations*: "In porn, everyone is comfortable with all sex acts that my partner may not be comfortable performing or simply not interested in."

- *Porn decreases interest in partners*: "I feel like porn makes my partner look 'ugly' in my eyes. I don't find my partner attractive anymore and my relationship of 20 months is suffering."
- *Porn causes insecurity*: "I feel like I can't make him as excited as the porn does, so I actually don't enjoy sex as much as I used to."

By far the most frequent comment on the topic of negative effects, however, was that there were none.

The educational value of porn may be greatest in places where school-based sex education is lacking or where it focuses more on "don'ts" than "do's." In Ireland, for example, sex-ed rarely includes depictions of genital or bodily diversity or a range of possible sexual positions. In a survey of Irish college students, most respondents reported obtaining from pornography basic information about bodily and genital anatomy and aesthetics, how to be a good sex partner, and how to explore diverse activities.[3]

Of course, porn is not designed to be educational, and some of what is "learned" from it—such as the average size of a man's penis or a woman's breasts or how long a man can thrust without ejaculating—may be misleading. In porn, women's orgasms are depicted much less frequently than men's, as if that's the natural order of things.[4] Condom use is spotty. And porn users get to choose what they view, so they don't learn much about aspects of sex that are at odds with their sexual orientations or interests.

DOES PORN HARM ACTORS?

Plenty of people think that making or watching pornography is bad simply because it's sinful. Porn is fornication on steroids, after all—something that is condemned in holy scriptures as well as the writings of St. Thomas Aquinas, the thirteenth-century ethicist whose views on nonmarital sex I mentioned in chapter 1. When porn is condemned simply by reference to religious authority, there is little that science can say about it.

Religious authority doesn't carry the same weight that it used to, however, so most modern-day critics focus on the harmful effects that they attribute to pornography. Porn is bad because it harms the actors engaged in porn production, because it legitimizes and encourages violence against women, or because consumers become addicted to porn and, in the process, lose the ability to function in their real-world sex lives. Such claims are appropriate topics for scientific inquiry: does porn really have those damaging effects?

Regarding harm to porn actors, one infamous case was that of Linda Lovelace, star of the 1972 mega-hit *Deep Throat*. Lovelace (whose real name was Linda Boreman) went on to become an antiporn campaigner. In books, speeches, and congressional testimony, she described her participation in *Deep Throat* and other pornographic films as involving physical coercion and abuse, much of it at the hands of her first husband. Other participants in the production of those films have variously confirmed and contradicted her allegations. Lovelace died in a car accident in 2002.

Porn production is far more professional today than it was in the 1970s, at least with respect to videos produced by the big-name U.S. companies. Still, even mainstream Hollywood has been plagued by abuse—it was the sex crimes of Oscar-winning producer Harvey Weinstein, after all, that set off the #MeToo social media campaign. It's hard to believe that porn production, which largely escapes media scrutiny, is free of those kinds of problems. In fact, several female porn actors have made allegations of nonconsensual sex acts by one leading male actor, James Deen—allegations that Deen has denied.[5]

Despite these concerns, many porn actors have spoken positively about their work. Here's one example: "I definitely think you need to be a very sexual person, because that's just the nature of the job. . . . I do love sex, but for me, that's not the point of doing porn. Porn, for me, is the ultimate fantasy. I like being on display, I like being on camera, I like turning people on, I like being a sex object. Being a porn star is the ultimate fantasy, to me."[6]

To the extent that porn actors make negative comments about their work, most are about minor or temporary physical problems:

> I cannot film more than 1 boy/girl scene in a week, because the sex is so intense and usually with a much-larger-than-average sized penis. My vagina needs a few days to recuperate. —Siri.[7]

> All that bouncing around can really hurt, especially if one of those silicone pumpkins accidentally whacks you in the head. —Anonymous.[8]

One group of researchers, led by Corita Grudzen, then at Mount Sinai School of Medicine, surveyed the mental health of female porn actors and a comparison group of other young Californian women.[9] They found that the porn actors were much more likely to report being depressed—33 percent did so—compared with 13 percent of the women in the comparison group. That doesn't accord well with the positive reports cited earlier. Still, the actors also had far more negative childhood experiences; many had been abused or grew up in poverty. Thus it's unclear whether their engagement in porn caused the women's depression or adverse childhood circumstances led to both their depression and their engagement in porn. Resolving this question would require a longitudinal study of women before, during, and after their acting career.

Porn production involves real sex and with it, the real risk of disease transmission. In some instances, porn actors have acquired HIV and other sexually transmitted infections. It's often unclear whether those infections were acquired in the studios or in the course of the actors' outside lives. One clear-cut example, however, involved twenty-one-year-old Lara Roxx, a newcomer to porn acting, who, in 2004, acquired HIV from actor Darren James.[10] James himself had apparently acquired the virus earlier in the course of making a film in Brazil. He tested negative at the time of the encounter with Roxx but tested positive later. (There's a lag between when an infected person becomes infectious to others and when they trigger a positive antibody test.) The James–Roxx incident led to a temporary halt in porn production.

Recent advances in testing procedures have greatly reduced the risk of HIV transmission in the industry. So has the development of drug treatments that suppress virus levels or protect against the acquisition of the virus. Still, some other STIs remain a problem. Condom use by porn actors is mandated in some jurisdictions, including Los Angeles, but some companies have evaded this requirement by moving production to less restrictive locales. Many consumers prefer to watch condomless sex, and there have been some efforts to remove condoms in postproduction using CGI technology. That's a questionable goal, though, because visible condom use in porn likely has some educational value, especially for teenage consumers.

Amateur porn and sexual webcamming (on-demand live porn) often feature solo sex, which largely sidesteps the problems of abuse and disease transmission. The main risk for webcammers is a failure to thrive financially: with the onset of the COVID-19 pandemic, many escort (off-street) prostitutes moved into webcamming. As a result, the market became over-saturated, and it's now next to impossible for newcomers to make a living. Another risk is that the videos will resurface years later, just as the actor has announced their candidacy for political office.

So there are pluses and minuses associated with acting in porn. A typical porn actor earns in the region of $1,000 for a day's work, though that varies considerably depending on the actor's popularity and willingness to engage in unconventional sex acts. For actors with limited earning potential outside the industry, that may well be enough to compensate for the risks inherent in the job.

DOES PORN HARM WOMEN?

Feminists of the 1970s represented all porn as harmful to women. In 1979, when Susan Brownmiller, Andrea Dworkin, Gloria Steinem, and Bella Abzug marched past the pornographic movie theaters of New York's Times Square, they did so behind a banner proclaiming that "Pornography is Violence Against Women." After the march, Dworkin gave a speech in which she said,

> The one message that is carried in all pornography all the time is this: She wants it; she wants to be beaten; she wants to be forced; she wants to be raped; she wants to be brutalized; she wants to be hurt. This is the premise, the first principle, of all pornography. She wants these despicable things done to her. She likes it. She likes to be hit and she likes to be hurt and she likes to be forced. Meanwhile, all across this country, women and young girls are being raped and brutalized and hurt.[11]

Dworkin must not have seen *Deep Throat*, because that film was humorous, lacked any kind of violence, and focused—at least superficially—on Linda Lovelace's sexual fulfilment. (Viewed today, it is also a nostalgic reminder of a time when porno-graphic films had plots and dialogue and Miami was a fun town to drive a Cadillac around in.) Even so, there was, and still is, plenty of porn that features simulated rape and other aggressive acts against women. Thus the question is, "Does watching such porn cause men to act out what they see?"

Efforts to address this question go back at least fifty years. In 1970, a commission set up by President Lyndon Johnson reported that pornography did little or no harm, but the report triggered angry rebuttals from legislators, religious leaders, and feminists. Two years later, *Deep Throat* brought porn out of the shadows, rebranding it as family fare—or at least as entertain-ment that liberal-minded couples could watch without shame. Porn seemed to be on its way to broader acceptance, but in 1986, a second government investigation, the Meese Report, con-cluded that violent pornography does cause violence against women and called for a national campaign to ban it.

Academic researchers have been engaged in this debate since the early 1970s. One of them is the social scientist Neil Mala-muth of UCLA. Malamuth has taken many approaches to the topic, including laboratory studies in which volunteers—often heterosexual male college students—are exposed to various kinds of sexually explicit material, after which their likelihood of acting aggressively toward women is assessed. His overall conclusion from his own and other studies is that most porn does not increase men's likelihood of harming women but that a small minority of men do react to violent pornography with heightened arousal and a heightened interest in sexual coercion. For these men, Malamuth believes, certain predisposing factors existed prior to their exposure to violent pornography. (I discuss this topic further in the following chapter.)

Experimental studies are scientifically rigorous—participants can be randomly assigned to view violent or nonviolent porn, for example—but they are also somewhat artificial. An alternative approach is to look for correlations between porn use and real-world sexual aggression. Convicted male sex offenders don't have a history of greater porn use than nonoffenders, according to a review of the literature by Emily Mellor and Simon Duff of the University of Nottingham in the UK.[12] In contrast, male high school and college students who say they watch violent porn are more likely to admit to having committed acts that meet the definition of sexual assault or rape compared with males who don't watch violent porn.[13]

But does that exposure to porn promote sexual violence, or does a certain developmental pathway lead to both those outcomes—a liking for porn and a proclivity to violence? To tackle this question, several researchers have undertaken longitudinal studies, asking whether viewing porn at time A predicts sexual violence at time B. The answer appears to be that it doesn't, according to a study of high school students led by Taylor Kohut,[14] and a study of college students led by Gabe Hatch of the University of Miami.[15] Such studies undercut the idea that there is a strong causal connection between porn use and sexual aggression.

It's even possible that viewing porn—including violent porn—has the opposite effect, reducing the viewer's likelihood of committing rape or sexual assault. In this way of thinking, porn provides an alternative outlet for men's sex drive that otherwise might show itself in sexual aggression. The effectiveness of such outlets has been demonstrated in real-life contexts. Thus in 2003, when a legal glitch rendered indoor prostitution legal in Rhode Island, rape offenses fell by 31 percent.[16] Similarly, whenever an adult entertainment establishment opens in New York City, the incidence of sex crimes in that locality drops by 13 percent, according to a study by the economists Riccardo Ciacci and Maria Sviatschi.[17] In some countries, the legalization of pornography has been followed by a drop in sex crimes, and in no country have sex crimes increased faster than other crimes after legalization.[18]

All in all, the evidence that porn promotes violence against women is weak. It's possible that some men who are already predisposed to commit sex crimes become more likely to do so after viewing violent porn. If they are, this influence may be counterbalanced or outweighed by porn's ability to provide an alternative outlet for sexually aggressive impulses.

It has frequently been claimed that aggressive or violent content in pornographic videos is on the increase—perhaps because men are becoming habituated to "vanilla" sex and therefore need ever more extreme stimulation to achieve orgasm. That's not the case, however, according to a study by the sociologists Eran Shor and Kimberly Seida of McGill University.[19] They examined popular videos uploaded to Pornhub from 2008 to 2016 and found no trend toward increasing violence: throughout the period, about four in ten videos included at least one instance of aggressive behavior, and about one of ten included aggressive behavior that was intended to appear nonconsensual.

One in ten amounts to a lot of videos, of course. In 2020, Pornhub removed about ten million user-uploaded videos from its site out of concerns about consent, violence, and actor age. There are other sites, however, that specialize in videos that show rape, other forms of violence, incest, or underage sex. While most of these illegal activities are simulated, some may be real.

DOES PORN HARM USERS?

"I used to watch a lot of porn, to be honest," pop singer Billie Eilish told Howard Stern in 2021. "I started watching porn when I was like 11. . . . I think it really destroyed my brain and I feel incredibly devastated that I was exposed to so much porn."[20]

For someone with a destroyed brain, Eilish functions surprisingly well in the world. But for some other people, most of them men, porn has serious adverse effects on their lives. They may say that viewing porn—usually accompanied by masturbation—steals time and energy from more important activities such as work or study, causes them to lose interest in sex with their regular partners, or leads to problems in sexual functioning, such as erectile dysfunction or difficulty reaching orgasm. They may try to compensate by watching more porn or more extreme porn or by adopting more vigorous or unusual techniques of self-stimulation. They may say they want to quit viewing porn or masturbating but are unable to do so. Among Americans who have ever viewed online porn, 7 percent of men and 2 percent of women "agree" or "strongly agree" with the statement, "I am addicted to pornography," according to a representative national survey led by the psychologist Joshua Grubbs of Bowling Green State University.[21]

The concept of porn addiction is controversial. Originally, addiction meant addiction to a substance, most commonly alcohol or opiate drugs. But psychiatrists and neuroscientists came to realize that some difficult-to-control behaviors are comparable to those of people with substance addiction.[22] Gambling disorder is the most widely accepted example; it's listed as a mental disorder in *DSM-5*. Similarities between gambling disorder and substance addiction include the short-term rewards of gambling, a need to gamble increasing amounts to achieve the same level of arousal, distress when unable to gamble, the use of gambling to cope with other life stresses, persistence in the face of damaging consequences, and unsuccessful efforts to quit. There are also similarities in terms of the neurotransmitters and brain systems that are involved, as well as the personality traits—impulsivity in particular—that predispose one to the disorder.

Some or all these symptoms can accompany other kinds of rewarding behavior—internet gaming, shopping, tanning, eating, and petty thievery (kleptomania)—which leads some experts to consider them to be behavioral addictions analogous to gambling disorder. Absent such symptoms, those behaviors are not addictions; that's most obvious in the case of eating, a vital life function.

If porn addiction exists, it's a subtype or component of sex addiction, a condition whose existence was popularized by the psychotherapist Patrick Carnes in his 2001 book, *Out of the Shadows.*[23] Sex addiction is not listed as a mental disorder in *DSM-5*, but the World Health Organization's 2018 listing of diseases, *International Classification of Diseases* (*ICD-11*), does have an entry for "compulsive sexual behavior." Although the symptoms of sex addiction are closely analogous to those of gambling addiction, a key difference is that sexual behavior is a normal life function. For that reason, some sex researchers consider that the concept of sex addiction is just a way of pathologizing forms or degrees of sexual expression that are unacceptable to mainstream society. Still, as with eating, it's not the activity by itself that constitutes a disorder but, rather, the harms caused to the users or others by their excessive focus on the activity.

While addiction to pornography got sporadic mention in the late twentieth century, it was the onset of free online video streaming—especially the founding of Pornhub in 2007—that opened the door to widespread problematic consumption.

Hundreds or thousands of men have described their porn addiction on sites such as reddit's r/NoFap. ("Fapping" is slang for male masturbation.) Porn addiction is largely self-diagnosed, and some clinicians and neuroscientists dispute whether the reported symptoms fulfill the accepted criteria for such a diagnosis.[24]

NoFappers' main problems with what they call "PMO" (porn, masturbation, and orgasm) are depression, damaged relationships, interference with other life tasks, constant craving, a need for more intense or prolonged stimulation, and difficulty achieving erection or orgasm. Some men want to quit porn and masturbation forever; others want to "reboot"—that is, quit long enough to regain a more normal level of sensitivity. Either way, these men find that quitting is difficult, and it often gets more difficult over time: "I don't know what to do anymore. I used to have 60–70 day [quitting] streaks, now I can't even get past 3 days without fapping. I just want to be free from this. I think there's a deeper issue at play here because for whatever reason in the past it was easier for me to resist PMO—but now it's so difficult."[25]

Most people who consume porn don't feel addicted; rather, they see porn as something enjoyable and positive in their lives, as suggested by the comments quoted at the beginning of the chapter. And according to one survey led by Beáta Bőthe of Université de Montréal, the frequency of porn consumption is not linked to sexual dysfunctions, only porn consumption that has addiction-like characteristics such as an increasing need for more extreme pornography and consumption in cycles of withdrawal and relapse.[26] So what is it about the minority that causes them to experience pornography in this fashion?

One idea, espoused by Joshua Grubbs among others, is that many men who suffer the most negative consequences of porn consumption, including the sense of being addicted, do so because their behavior is inconsistent with their moral beliefs. They experience an inner conflict—what Grubbs terms a "MORAL INCONGRUENCE"—and this leads to guilt about the behavior and a ceaseless struggle between indulgence and abstinence. Grubbs and his colleagues have shown that there is indeed a correlation between moral disapproval of pornography—especially disapproval based in religious beliefs—and problematic consumption of porn and the sense of being addicted.[27] This perspective suggests that exploration of this inner conflict in therapy, and the encouragement of more liberal views in the area of sex, might help alleviate porn addiction. That would parallel the use of gay-affirmative therapy in the treatment of persons conflicted about their same-sex attractions.

Yet there are also many men who don't fit the moral incongruence model. An example is Alexander Rhodes, who is considered the founder of the NoFap movement and runs the NoFap.com website. Rhodes says he became addicted to porn during his teens; by his college years he was watching porn up to fourteen times a day. He could maintain an erection during sex with his girlfriend only by thinking about porn. "I'm very sex-positive," Rhodes told the *New York Times* in 2016. "I'm not a religious person. I'm not someone who supports religion. I'm not against religion, but I don't support it. And I completely, firmly believe in premarital sex."[28]

Men who consider themselves addicts without being morally conflicted may have personality traits that predispose them to addiction, such as impulsiveness and novelty-seeking, or gene variants that are known to be associated with addiction, or they may have been exposed to social stresses that make addictive behaviors more likely. But given how extraordinarily powerful a stimulus online porn can be—the high-definition close-ups, the hyper-attractive actors, the endless new faces and wilder antics—it's more of a question why most users don't become hooked rather than why a minority do.

Although men are the overwhelming majority of those who post online about their porn addiction, some women do also. Even though these women are as badly affected as the men, their low numbers tend to undermine their credibility. Also, many women are less focused on videos and are more into narrative porn, which men may not consider porn at all. Here's one comment on r/NoFap:

> Society tells us women that fapping is an addiction only men have. So if you're a woman who faps 20 times a day, blows off dinner with the girlfriends for solo night fapping and whose closest friend is your vibrator, society says, "Wow, you're so sexually enlightened and sex-positive!" . . . By the time many women have gotten to the point where they say "I think I have a sex addiction," things have gotten really bad. . . . At my worst, I would have 7 or 8 different internet pages and go through them for about 3 or 4 hours or more, looking for the perfect sex story to get off on.[29]

Thus, although I made a flippant comment about Billie Eilish at the beginning of this section, there's no reason to dismiss her story, even if she expressed it in an over-the-top way. She deserves credit for escaping from the trauma that heavy porn use caused her.

Another issue concerns pornography consumption by adolescent or preadolescent minors. It's been suggested that exposure to porn (and sexual topics on television) leads to "premature sexualization," with attendant risks such as exposure to sexually transmitted diseases, teen pregnancy, and mental health problems. It is the case that girls and boys who consume porn early and in relatively large amounts become sexually active earlier than those who consume little or no porn, in terms of both solitary masturbation and partnered sex.[30] What's not clear is whether early porn consumption causes early sexual initiation or whether both behaviors result from a common cause, such as early physical maturation. Regardless of the direction of causation, what's needed is early and comprehensive sex education, which helps adolescents make informed choices about sexual initiation and avoid its possible adverse consequences.

The bottom line is this: most people derive pleasure and other benefits from porn, but some do suffer adverse consequences, and a few become trapped in a state that deserves to be called a behavioral addiction just as much as gambling does. Helping such people escape from their situation requires understanding of what led each particular person into their problematic use of pornography, as well as peer support on their route to recovery.

CONCLUSIONS

In this chapter, I've focused not so much on the technology of pornography, which is constantly evolving, or on porn's content or style, which also changes over the decades, as on the possible benefits and harms of porn production and consumption. Among the benefits reported by porn users are the sheer pleasure of watching it, as well as its ability to bolster sexual arousal, improve communication between partners, educate, promote experimentation, provide an alternative outlet, and reduce guilt. Some of these reported benefits are limited or questionable—the educational information provided by porn can be quite misleading, for example. Still, the perceived benefits of porn are attested by the sheer volume of consumption.

The supposed harms of porn include some that are alleged by outsiders and others that porn consumers themselves experience. Critics outside the porn industry allege that porn harms actors, degrades and objectifies women, and encourages sexual violence. That last allegation is the most serious one, and it does get some support from the work of Malamuth and others. Still, the violence-encouraging effect seems to be a small one, affecting only certain men who watch certain kinds of porn. Looked at through the lens of nation-by-nation sex crime statistics, there doesn't seem to be a meaningful effect, and there may even be a countervailing effect by which porn consumption diverts the potential for sexually aggressive behavior away from the real world.

The ill-effects experienced by some porn users—including an increasing dependency on porn and difficulties with relationships and sexual performance—seem to be real, to judge by the large number of reports by members of online groups such as NoFap. These difficulties may be attributable to moral incongruence, the conflict between a person's moral beliefs and their actual behavior.

Still, it would be surprising if porn consumption didn't have some general effect of this kind, regardless of moral beliefs. After all, there's a general effect in psychology whereby exposure to strong stimuli impairs responses to weaker ones. It's been shown experimentally, for example, that viewing images of very attractive women causes married heterosexual men to down rate the attractiveness of "average" women and even to rate themselves as less in love with their wives.[31] Whether contrast effects of this kind can cause durable problems of the kind reported by NoFappers is uncertain but plausible.

Whether or not porn consumption encourages sexual violence against women, rape is an enormous social problem. I therefore devote the following chapter to this, the darkest expression of human sexuality.

10

RAPE

In 2003, an international group of population geneticists
led by Chris Tyler-Smith of Oxford University announced
a remarkable finding. They had sampled Y chromosome
DNA from more than two thousand men belonging to twenty-
six ethnic groups in Asia. From their analysis of this DNA, the
researchers concluded that one in twelve of all the men who live
in a broad swath of Central Asia between the Caspian Sea and
the Pacific Ocean—about 16 million of the 190 million men in
this region—belong to a single lineage that originated in Mongolia about one thousand years ago. This lineage, the researchers
proposed, included Genghis Khan, the founder of the Mongol
empire, and his male relatives.[1] (More recent work suggests that
the lineage identified by Tyler-Smith included not just the ruling clan but also many of the common Mongol warriors, who
may have had ancestors in common with the Khans.[2])

The Mongol rulers, as well as their warriors, were men of
an extraordinarily violent disposition.[3] Genghis himself, when
he was only thirteen, got into a dispute with his brother over a
bird and a minnow that he had caught and shot him to death
with arrows. By the end of his life a half-century later, Genghis was responsible for the deaths of uncountable numbers of

people, most of them the inhabitants of cities captured by the Mongols.

Often, when cities were captured, the women were spared—spared death, that is. Instead, they were raped and taken as concubines or sold into slavery. Always, by Mongol custom, the most beautiful women were offered to the leaders for their personal use—to Genghis, his sons, grandsons, or whichever of his male relatives were in the field on that particular campaign.

Small wonder, then, that the Khans now have sixteen million descendants—and that's only counting the males and only males who are descended from the Khans through an uninterrupted male line. It is likely that at least a drop or two of Genghis's blood run in the veins of most present-day Asians, male or female.

Yet if Genghis Khan was a man on a mission to spread his genes, he seems to have made one serious error. In their DNA study, Tyler-Smith's group tested men over a wide area of Asia, but they didn't test men in Russia even though a large part of what is now Russia once lay within the Mongol empire. (This portion of the empire was known as the Golden Horde.) That deficiency was made up by a group of Russian and Polish geneticists, whose results were made public in 2007.[4] They came up with a shocker: aside from ethnic Mongolians and groups living close to Mongolia, Genghis Khan's Y chromosome isn't present in Russia at all.

How could it be that the Khans spread their genes all over the rest of the Mongol empire but failed to do so in Russia? The scientists didn't put forward any explanation, but a careful reading of the history books offers one. The Golden Horde was founded and ruled by Batu, son of Genghis's eldest son, Jochi. Some unusual events surrounded Jochi's birth—events that are retold in the *Secret History of the Mongols*, an anonymous account that dates in part to the time of Genghis and was probably influenced or approved by the emperor himself.[5]

Genghis married his primary wife, Börte, when he was sixteen. A few years later, before the couple had any children, Börte was abducted by the Merkits, a tribe of Mongols who lived far to the north. (They abducted her in revenge for a similar crime that had been perpetrated in the reverse direction a generation earlier: Genghis's father had abducted a Merkit girl.) When the Merkit raiders returned to their homeland, they handed Börte over to a man named Chilger Bote, the younger brother of the Merkit man, now deceased, whose wife had been stolen earlier. A kind of rough justice, if you like, especially for a culture that viewed women as property.

To revenge himself in turn for this insolent abduction, Genghis organized an army of twelve thousand men that crossed the mountains and attacked the Merkits, who were encamped to the south of Lake Baikal. The Merkits fled in disarray, and Genghis was able to recover Börte, along with a large number of Merkit women, who served as rewards for Genghis's helpers. Börte returned to her wifely duties, and a few months later, the pair's first son, Jochi, was born.

A few months later? How many months, exactly? We are not told, but we *are* told that there was widespread suspicion among the Mongols concerning Jochi's paternity. Genghis insisted that Jochi was his son and treated him as such, but so great was the suspicion of Jochi's illegitimacy that years later, when Genghis was Great Khan and ready to nominate his successor, he was not allowed to choose Jochi. Instead, he chose one of his younger sons, Ögedei.

Genghis did give Jochi sovereignty over Russia, but Jochi died early, and the Golden Horde was ruled by Jochi's son Batu, who was enormously successful in expanding the empire. He followed or outdid Genghis's example in terms of murder, rape, and pillage. Thus Batu very likely did leave his genetic mark on the Russian and Eastern European territory he conquered; but it wasn't Genghis's mark, because evidently Batu wasn't Genghis's biological grandson but the grandson of a Merkit nonentity called Chilger Bote. Thus Chilger became a biological father of the Russian nation without ever having to leave home, simply by virtue of having once impregnated the wife of the future Mongol emperor.

Genghis Khan's Y chromosome is not the only example of its kind. Giocangga was a sixteenth-century Manchu warrior and patriarch of the Qing dynasty (the last Chinese dynasty), whose rulers demonstrated their wealth and power by taking innumerable wives and concubines. A type of Y chromosome possessed by 1.6 million present-day Chinese men can likely be traced back to Giocangga, according to a 2005 study by Tyler-Smith's group.[6] The Frankish and Ottoman empires also seem to have left their genetic imprints in present-day populations.

When thinking about these findings, two alternative possibilities come to mind. One is that there was something biologically special about the Khans' Y chromosome that was adaptive in evolutionary terms—that is, something that helped its carriers have more offspring. For example, the chromosome might have carried genes that conferred a heftier-than-usual dose of physical and sexual aggressiveness, promoting the very behaviors that allowed the Khans to replace large segments of the population with their own descendants.

The alternative possibility is that Genghis Khan's Y chromosome was a so-called adaptively neutral variant, meaning that it neither promoted nor impaired its owners' reproductive success. In this scenario, the Khans were aggressive and psychosexually overendowed for reasons that had nothing to do with their

Y chromosomes. The real reason lay in genes located on other chromosomes or in entirely nongenetic influences, such as the fact that the Khans were brought up to value physical and sexual aggression and had ample opportunity to put these traits into practice. Their Y chromosome simply benefited from this behavior: it got an extraordinary free boost into the future.

Tyler-Smith calls this latter mechanism "social selection," and he believes that it is the correct explanation for the prevalence of Genghis's Y chromosome today. The reason he gives for this conclusion is that, as discussed in chapter 1, the Y chromosome carries only a small number of genes, and those have specialized functions that seem unlikely to promote the behaviors for which the Mongol rulers were infamous. Still, the question is not yet resolved: recent evidence does suggest a role for certain regions of the human Y chromosome in aggressive behavior.[7]

Regardless of the exact mechanism, it seems clear that historically, a disposition to violence and nonconsensual sex has helped many men propagate their genes. This disposition helped their owners reproduce, but at a tragic cost for society as a whole.

The traits of physical aggressiveness and interest in casual, promiscuous, or coercive sex differ very significantly between the sexes. In the contemporary United States, men are nearly ten times more likely than women to commit murder and nearly fifty times more likely to commit sexual assault. What's more, the gendered nature of these traits exists across all cultures and historical periods that have been studied, and (with regard to aggressiveness) it emerges early in childhood.[8] If you come across a study that reports equal aggressiveness by men and women, you will find that it does so by including nonphysical or indirect forms of hostility, such as malicious gossip.

Why are men more aggressive than women, physically and sexually? Radical feminists have traditionally laid the blame on the different ways in which boys and girls are brought up and the unconscious lessons they absorb from their families, television, and so on, especially with regard to sexually aggressive behavior. If you do an internet search on phrases like "boys are taught to" or "boys are socialized to," you will find hundreds of propositions of that kind. The late sociologist Diana Russell, for example, wrote in 1984, "Males are trained from childhood to separate sexual desire from caring, respecting, liking or loving. One of the consequences of this training is that many men regard women as sexual objects, rather than as full human beings. . . . [This view] predisposes men to rape."[9]

In fact, radical feminist thinking has traditionally downplayed—or outright denied—any sexual motivation for rape, asserting instead that it's about violence as a means of control. It's about the control of a specific woman by a specific man, as well as the control of all women by all men—the patriarchy, in a word.

More recent feminist thinking has taken a more nuanced view of the matter. Beverly McPhail, for example, who, prior to her retirement, ran the Women's Resource Center at the University of Houston, has presented a multifactorial model for the causation of sexual violence. In her model, sexual gratification, revenge, recreation, power/control, the desire to achieve or demonstrate masculinity, and issues of race and class may all be combined to varying degrees.[10] Yet such ideas are often expressed in complex academic discourse. They cannot compete with "rape is not sexual," a four-word mantra that appears on the website of the Valley Crisis Center of California's Merced County, as well as on countless similar sites.

One virtue of multifactorial models is that they potentially account for something that is largely ignored in the earlier patriarchy-linked thinking: that only some men commit rape. Thus Neil Malamuth of UCLA, whose work on the effects of pornography I discussed in the previous chapter, has put forward (in collaboration with Gert Martin Hald) a "confluence model," which posits that three developmental threads must coexist for sexual aggression to occur.[11] One thread begins with an abusive home environment, leading to early delinquent behavior, early sexual initiation, and sexual promiscuity. A second thread begins with a hostile personality, leading to narcissism and the tendency to be aroused by sexual dominance. The third thread is a failure to develop the trait of empathy, because empathy, in Malamuth's studies, blocked the expression of sexual aggression even when all other factors promoted it.

The proponents of such multifactorial models seldom place much emphasis on biological and evolutionary factors as helping to explain why some men commit rape. This reluctance may be due to a fear that those approaches potentially excuse or even justify rape. But they do not: rape is a grave crime no matter what its causes may be.

Among the biological factors contributing to the sex difference in sexual aggression, one has received most attention: testosterone. Serum testosterone levels are about tenfold higher in healthy men than in healthy women, with essentially no overlap between the sexes. Castration of males (removal of the testes, thus eliminating the principal source of testosterone) decreases or extinguishes their sex drive and makes them less aggressive. This is true in both humans and many animals.

Still, the effects of castration in humans take some time to show themselves—sometimes months or years—even though the reduction in testosterone levels is nearly instantaneous. This suggests that testosterone does not regulate the sex drive or aggressiveness on a short-term basis. In fact, a healthy man's blood testosterone levels fluctuate wildly over the course of the day, but those fluctuations do not translate into equivalent hour-by-hour fluctuations in sexual or aggressive feelings or behavior. Rather, testosterone appears to play a long-term role in maintaining the activity of brain circuits that are involved in generating these feelings and behaviors.

What about differences in testosterone levels within the male population? Among men, high testosterone levels are correlated with an increased likelihood of engaging in criminal violence and other forms of aggression; male aggression and sexual coercion are most prevalent during young adulthood, when testosterone levels are high. But the correlation between testosterone levels and aggression is not especially strong; it leaves most of the variation in aggressiveness unexplained.[12] In fact, one meta-analysis that focused specifically on men convicted of rape or other forms of sexual aggression failed to find an increased testosterone levels in those men, compared with men convicted of nonsexual offenses or nonoffenders.[13]

Another causal factor may be the levels of testosterone during fetal life, when the brain circuits involved in aggression and sexuality are first assembling themselves. This statement is based on indirect evidence, such as the measurement of anatomical details—ratios of the lengths of different fingers, in particular—that are believed to reflect testosterone levels during fetal life, as discussed in chapter 4. Men with finger length ratios indicative of high fetal testosterone levels are significantly more aggressive, on average, than men whose ratios are indicative of low fetal levels. Still, as with the adult testosterone studies, the correlation between finger length ratios and aggressiveness is not very strong.[14]

Other biological factors are also likely to play a role. The brain neurotransmitter serotonin, for example, appears to help regulate aggressive behavior, especially impulsively aggressive acts like Genghis Khan's killing of his brother.[15] Violent criminals tend to have low serotonin levels, and increasing serotonin levels with drugs tends to reduce aggressive behavior, at least among mentally ill individuals.

Genes influence aggressive behavior. Based on twin studies, it appears that genes account for about half the total variability in physical aggressiveness among men. (That is a fairly typical level of heritability for psychological traits in general.) For the most part, the actual genes involved have not been identified. Dutch researchers achieved an important breakthrough in 1993, however, when they studied a family in which nine male members showed severe impulsive aggression, including assault, attempted rape, and arson.[16] The men were easily provoked into outbursts of rage, cursing, or violence by stressful events that most of us would cope with more calmly. The researchers showed that the affected men (but not the psychologically normal men in the same family) had a nonfunctional gene for an enzyme named MONOAMINE OXIDASE TYPE A (MAOA). This enzyme breaks down certain neurotransmitters or neuromodulators: serotonin, norepinephrine (noradrenaline), and dopamine. Male mice in which this gene has been knocked out also show hyperaggressive behavior, a finding that strengthens the causal link between the gene and aggressiveness in the Dutch family.[17]

Fully nonfunctional MAOA genes, as in the Dutch family, are rare; more common are variants of the gene that are less active than normal. These variants are associated with an increased risk of violent behavior by adolescents and young adults, especially in males who have been exposed to childhood trauma.[18]

Aggression has survival value, and that's why evolution has ensured that humans have aggressive feelings and the capacity to show them in aggressive behavior. Aggression can supply all kinds of resources, ranging from food and living space to mates, as we saw with Genghis Khan. Genghis Khan and his male relatives left at least sixteen million present-day descendants. Milquetoast Khan (a creature of my imagination) left none.

But in evolutionary terms, aggressiveness has more value for males than females. That's because aggressive behavior is risky. It's a high-stakes game that offers the lure of great rewards but also the threat of catastrophic losses, including premature death. This kind of gambling usually makes more sense for males than females, on account of differences between the sexes in the amount of investment they put into reproduction—a topic explored by evolutionary biologist Robert Trivers in the 1970s.[19]

Females of most mammalian species make a much greater investment in reproduction than males do. To reproduce, a female has to invest all the time and metabolic expense of pregnancy, followed by lactation and infant care, before she can think of producing more offspring. Thus females are limited in terms of the number of offspring they can produce over a lifetime, but they generally do get fairly close to that limit.

A male, on the other hand, just has to inseminate a female with a few drops of semen, and then he's free to head off in search of another mate. So males can have much larger numbers of offspring, but—because of choosiness by females and competition from other males—they may very easily have none. In other words, there is more variability in reproductive success for males than for females. The sex difference in variability is what offers a greater incentive for risk taking by males than by females. To put it simply, why should females take risks when they can get pregnant and have offspring without doing so?

There are all kinds of complications and caveats surrounding this idea, of course. For one thing, it isn't always the females who make the larger investment in reproduction. In a few species, males do most of the hard work. There are waterbirds called jacanas, for example, in which the males incubate the eggs and feed the nestlings, leaving the female free to fly off and mate with another male. In such species, however, it's the females who are the aggressive sex, competing fiercely for access to males. In other words, these exceptions validate a more general rule: the sex that gets off lightly in terms of parental investment is the one that's more aggressively competitive in its reproductive strategy.

In humans, that's the males. However, the asymmetry between the sexes in this respect is not as great as it is in some other mammals. Men do typically put some investment into reproduction, such as providing food, resources, and protection for the family. That's in contrast to many other species in which the male does absolutely nothing aside from inseminating the female. Part of the reason men put some work into reproduction is that they increase their children's chances of survival; human infants require more and longer parental care than the young of most other mammals, after all. In addition, women probably choose partners who exhibit a commitment to the family in one way or another.

So far I have basically treated male aggression as a single coherent trait whose benefit is to maximize an individual's reproductive success. Genes promoting Genghis Khan–type behavior have evolved because of the positive effect they can have on their owner's reproduction and hence on the survival of those genes. This is the "selfish gene" perspective made famous by Richard Dawkins.

Of course, some forms of aggression are more closely tied to reproduction than others. Abducting women to stock your harem has direct reproductive significance. Seizing someone's land doesn't—it could be seen as an economic form of aggression that has to do simply with surviving or living better than you would have done otherwise. Some opponents of the selfish gene perspective, such as the evolutionary biologist Niles

Eldredge, have argued that most behavior is "economic" in motivation rather than reproductive—including a great deal of behavior that is overtly sexual. He said, for example, that much heterosexual intercourse is motivated not by the urge to make babies but by a deal between partners in which sexual pleasure is traded for some kind of resource.[20] Prostitution is the extreme case of such a deal.

There are a couple of problems with Eldredge's point of view. For one thing, acknowledging the economic motivation of much human behavior doesn't really undermine the selfish gene perspective, since it's necessary to live in order to have children. Living well (having lots of economic resources and power) allows one to have more children than one would otherwise. So from a selfish gene perspective, economic strategies can be viewed as reproductive strategies, just less direct ones.

It's true, as Eldredge stressed, that rich people today don't have more children than poor people. In fact, they have fewer children, suggesting that wealth accumulation lacks any ultimate reproductive purpose. But Eldredge's mistake was to assume that evolution must give people the conscious urge to have children in order to get them to engage in reproductive behavior. Of course, that's not true. Evolution didn't even bother to tell people that sex causes pregnancy, a prerequisite for that argument to work. Yes, nearly everyone understands that causal connection today, but the further back we go in human evolution, the less likely it is that people grasped it.

What evolution did do is give people the urge to have sex, the most desirable form of which (for most but not all people) is penile-vaginal intercourse. If people engage in this behavior, the raw biology—ejaculation, sperm migration, fertilization, implantation, and pregnancy—will take over from there, with no further urges required. Babies will be born, and there are then other mechanisms, especially parent-child bonding, that take over to help the babies survive. If people happen to realize the causal connection between coitus and pregnancy and develop the skills (withdrawal, rhythm method) or technology (condoms, pills) to engage in the former without triggering the latter, they can and do happily subvert the original adaptive purpose of sex, getting the reward without paying the price. I discussed that topic in chapter 1.

Economic considerations are often relevant to sex and reproduction. Many sexual relationships that no one would equate with prostitution nevertheless have the quality of a deal—my beauty for your wealth, for example. And when industrialization reduces the economic value of children, couples have fewer of them; that is the "demographic transition" that is reining in population growth worldwide. But, contrary to a purely economic theory of sexuality, people don't have fewer children by engaging in less sex: they use contraception, including nontechnological methods such as withdrawal.[21]

What does evolutionary psychology have to say about that ultimate meeting point of sexuality and aggression: rape? Randy Thornhill (University of New Mexico) and Craig Palmer (University of Colorado) addressed this issue in a controversial 2000 book.[22] Thornhill's prior research was on mating behavior in insects—specifically, the *Panorpa* genus of scorpionflies. He showed that forced copulation is an adaptive strategy used by male flies; they even have a special organ that is used only to grasp a resisting female during copulation. However, males engage in forced copulation only when their preferred strategy—luring the female into voluntary copulation by means of a nuptial gift—is not possible because they can't find a gift. Provide such a male with a suitable gift and he will immediately switch strategies.

Forced copulation is commonly seen in an animal much more closely related to ourselves, the orangutan.[23] Fully mature male orangutans have cheek flanges and throat sacs that allow them to make "long calls," which attract females and warn off other males. But males may spend many years after puberty before they develop flanges and sacs—in fact, they may never do so. These subadult, unflanged males cannot make long calls, yet they are highly motivated to have sex with females. Although females sometimes accept sexual advances from unflanged males, they commonly refuse them. In such cases, unflanged males may resort to force—attempts at copulation which the female very obviously resists, sometimes leading to injury to one or both parties. Because unflanged males continue to grow after puberty, whereas females do not, they eventually become stronger than females. In addition, two unflanged males may cooperate in overcoming the female's resistance.

Thornhill and Palmer considered whether something similar may be true for humans. They cited research by Malamuth that suggests that men who lack resources, are socially disenfranchised, or are unable to gain access to sex partners for a variety of reasons are more likely than other men to adopt coercive sexual strategies.[24] These conditions are particularly effective triggering factors when they operate early in life. (It's interesting in this context that Genghis Khan grew up in conditions of considerable deprivation after the early death of his father.)

Although Thornhill and Palmer saw parallels between forced copulation in insects and rape in humans, they did not conclude that rape is necessarily an evolutionary adaptation— even though almost everyone who reviewed or commented on the book, including Eldredge, stated that they did draw that conclusion. Rather, they wrote that the available evidence did not allow them to decide whether rape is an adaptation or merely a by-product of selection for some other trait, in the same way that the redness of blood is a by-product of selection for blood's oxygen-carrying capacity. What Thornhill and Palmer did conclude was that, whatever the exact mechanism, the human capacity for rape has been shaped by evolutionary processes and is not simply a cultural phenomenon that can be understood only in terms of the dynamics of contemporary society.

One attempt to mathematically model the evolutionary costs and benefits of rape (in an American Indian population in Paraguay) concluded that the reproductive benefits of rape to the rapist are generally outweighed by the reproductive costs—in other words, he would leave more offspring, on average, if he refrained from rape.[25] Nevertheless, that model predicted that rape would serve the interests of the rapist—it would increase his total reproductive success—if some factor severely decreased his likelihood of producing offspring by consensual sex. This factor could be a lack of resources (as with the scorpionflies) or a lack of attractiveness to the other sex (as with the unflanged male orangutans).

In summary, evolution doesn't necessarily cause all members of a species to act nicely to one another. Rather, it can lead to intense conflicts, including conflicts between the interests of men and women, with sometimes tragic consequences, such as rape. The rejection or even "honor killing" of rape victims by their husbands or families—vile crimes though they seem to us—have a horrible logic in evolutionary terms. If Genghis Khan had rejected or killed his wife after her involuntary stay with Chilger Bote—actions that assuredly would have been acceptable by the standards of the time—his Y chromosome might be even more widespread today than it is.

RAPE AND THE BRAIN

For obvious reasons, we cannot record what happens in the human brain when a person commits rape, but some observations in mice hint at just how closely sexual and aggressive behaviors can be linked in terms of the neural activity that underlies them. In chapter 3, I described studies by Dayu Lin's team at New York University, in which they focused on the hypothalamus in male mice. When the researchers activated neurons in the medial preoptic area (MPA) of the hypothalamus, the mice would approach females and, if the females cooperated, mount and copulate with them.

What I didn't mention in that chapter was what happened if the females didn't cooperate. The males' behavior in that situation depended on how strongly the MPA neurons were activated. With low levels of activation—perhaps corresponding to levels typically seen in a natural situation—some males did nothing, while others made futile attempts at sex, achieving only shallow penetration at best, before desisting. When the researchers increased the activity of MPA neurons, however, all the males attempted to copulate, and after several unsuccessful attempts, half the males attacked the females. This was not simply a general increase in aggressiveness, because those frustrated males were no more likely to attack other males than they would have been if the MPA neurons had not been activated. It was a specifically sexual aggression.

This may not be a close parallel to human sexual assault, because the aggressive behavior by the frustrated male mice did not appear to serve the males' sexual purposes: the females didn't become cooperative in response to the male's attacks, as far as I could tell from Lin's brief account. Without that cooperation, penetration is well-nigh impossible in mice. Still, the observation that a "friendly" advance can so easily be transformed into an outright attack just by ramping up the activity of a tiny cell group in the hypothalamus hints at the disturbingly close relationship between lust and loathing—in the minds of mice, at least.

PREVENTING RAPE

Between the 1970s and 2013, the reported incidence of rape in the United States fell steadily and profoundly. The reason for the decrease is not known with any certainty, but it outpaced the decline in other violent crimes, and for that reason it can't be attributed solely to broadly "civilizing" factors such as the decreased use of crack cocaine and other drugs. Feminist activism and the resulting increased attention to the issue by men, women, jurists, and politicians were likely key factors.

Starting in 2013, the reported incidence of rape increased again. In part this was simply because government statisticians redefined rape in a more inclusive fashion. That was probably not the complete story, however, because the reported incidence of rape also increased in some other countries during the same period. In England and Wales, it increased an astonishing fivefold from 2013 to 2018, for reasons unknown.[26] Then, in 2020 and 2021, the reported incidences of rape fell again in both the United States and UK, most likely as a result of social isolation during the COVID-19 pandemic.

I stress *reported* incidence because it's believed that most rapes are not reported. Thus it's not entirely clear how much the actual frequency of rapes has changed in recent times. Still, the fluctuating numbers have led some critics to dismiss the evo-psych approach to rape; if rape is a product of evolution, it's been argued, its frequency should change over millennia or hundreds of millennia, not over the course of a single lifetime.

According to Thornhill and Palmer, however, the evolved proclivity to commit rape is context dependent. If the cost of rape goes up—due to a greater likelihood of detection or more severe punishments—the incidence of rape will go down. Conversely, if the cost of rape goes down, as may happen during Mongol-style military campaigns, then the incidence of rape will increase. These costs can change rapidly as, for example, during a police strike.

In their 2000 book, Thornhill and Palmer detailed how they believed a greater understanding of the evolutionary and biological basis of rape could lead to better strategies for the prevention of rape. Most of what they wrote failed to support their argument, however. They explained how the imprisonment of rapists would decrease the incidence of rape, for example, but that's surely true regardless of what the causes of rape may be.

Thornhill and Palmer did make one interesting point. If rape is motivated by the sex drive more than by hostility toward women, then treating rapists with drugs that reduce the sex drive—ANTIANDROGENS—should reduce the likelihood of repeat offenses. If, on the other hand, rape is not primarily sexual in motivation, there might be less reason to expect antiandrogens to have such an effect.

This distinction doesn't offer a foolproof test of the sexual versus nonsexual cause of rape, because lowering testosterone levels may decrease aggressive behavior in general, not just sexual aggression. Still, it's worth asking whether convicted rapists or other sex offenders who take antiandrogens are less likely to commit another sex offense than those who do not take such drugs. Unfortunately, only a handful of studies have addressed this question since the time of Thornhill and Palmer's book.

The findings, while suggestive, are not robust enough to conclude that antiandrogens offer an effective means to prevent reoffending by convicted sex offenders.[27]

Recently Mark Huppin and colleagues at UCLA presented another set of recommendations for rape prevention that were said to be based on evolutionary thinking.[28] Again, though, most of the recommendations—for example, the idea that extended interventions are more likely to be effective than brief ones, that there should be an emphasis on female empowerment, or that there should be clear antirape laws that instill a real fear of punishment—could be put forward by anyone concerned with preventing rape, even by someone who doesn't believe that evolution happened at all. In other words, the relevance of evo-psych theories to the real-world challenge of preventing rape, though a topic worth pursuing, remains uncertain.

HEALING THE TRAUMA OF RAPE

Women, men, or children who experience sexual assault—whether it be rape, attempted rape, or assault without penetration—are liable to suffer both immediate and long-lasting harms. Here I focus on the set of long-lasting psychological harms known collectively as posttraumatic stress disorder (PTSD), which in the context of sexual assault may be called RAPE TRAUMA SYNDROME. The symptoms of rape trauma syndrome may include flashbacks, anxiety, depression, lack of concentration, insomnia, sexual disorders, eating disorders, and social isolation, though the specific symptoms vary from person to person. In the case of childhood sexual victimization, the long-term harms can also include schizophrenia, self-harm, and substance abuse.[29]

Rape trauma syndrome may abate over time without specific therapy, especially if the traumatized person enjoys sensitive supportive from their partner, family, or peer group. Yet, as with PTSD caused by war experiences, the trauma may persist or worsen, even to the point of inducing suicidal thoughts or actions.

Many current approaches to the alleviation of rape trauma syndrome include exposure therapy, a treatment modality pioneered by the clinical psychologist Edna Foa and her colleagues at the University of Pennsylvania.[30] In Foa's conception, PTSD is caused by the development of a cognitive structure in which memories of the triggering event, emotional responses to that event such as fear, as well as unrealistic beliefs about the meaning of those responses become so tightly interlinked that irrelevant experiences or thoughts readily trigger the entire structure. In exposure therapy for rape trauma syndrome, the person is encouraged to recall the assault and the surrounding events within a safe therapeutic environment, allowing for the development of a competing cognitive structure in which memories of the event become linked to nontraumatizing memories and thought processes.

The efficacy of exposure therapy for various forms of PTSD, including rape trauma syndrome, has been validated in clinical trials. In a typical trial, persons suffering from PTSD are assigned randomly to exposure therapy or to a waiting list to receive that treatment at a later time. It has been consistently reported that the treated participants improve over the course of their treatment program, whereas the waitlisted participants show little or no improvement over that same period.

Despite this good track record, there are problems with exposure therapy. First, not all participants improve. Second, the treatment regime is demanding for both the participants and their providers: A typical treatment involves eight to twelve sessions spread over a few weeks or months, each lasting ninety minutes, along with daily homework. That makes exposure therapy costly, and its availability often fails to meet the demand. Also, significant numbers of participants drop out without completing the program, and these dropouts are often lost to follow-up. Therefore, outside the rigorous monitoring conditions of clinical trials, the overall value of offering exposure therapy is uncertain.

An alternative form of therapy known as LIFESPAN INTEGRATION was developed by Peggy Pace, a psychotherapist in private practice in Cle Elum, Washington.[31] Pace has proposed that a person suffering from PTSD perceives the traumatic event as having occurred more recently than it actually did. The person has, as it were, telescoped the time that has elapsed between the event and the present, making the event more salient than it should be. A major goal of lifespan integration is to help the person gain a more realistic sense of how much time has elapsed. To achieve this, the person is asked to construct a sequence of intervening events and go over these events one by one in therapy. This procedure, Pace believes, helps distance the person from their trauma, reanchoring the trauma in the past and thus making it feel less immediate and less threatening.

In 2020, a Swedish research group published the findings of a controlled trial of lifespan integration therapy conducted at a specialist clinic for the treatment of sexual trauma.[32] Women who had suffered a single sexual assault within the five years prior to enrollment and who were diagnosed with PTSD were randomly allocated to the treatment program or a waitlist. Prior to treatment, each woman was asked to list a couple of dozen memory cues—phrases such as "yellow bicycle"—that brought to mind a specific, vividly recalled event in her life subsequent to the assault. Collectively, the sequence of memory cues covered the entire time span between the sexual assault and time of treatment. Then the woman had a single, approximately two-hour session with the psychotherapist during which she described each event and then was cued by the therapist to jump sequentially from the earliest to the most recent event. In essence, this procedure was intended to push the sexual trauma backward in time by expanding the woman's sense of the duration of the intervening period. The findings were very positive: 72 percent of the women in the treatment arm no longer fulfilled a diagnosis of PTSD after treatment, compared with 6 percent of the women in the comparison arm, and the benefit persisted for at least six months. Even those women who were not "cured" experienced a lessening of their symptoms.

If these results can be replicated by other groups, lifespan integration therapy should offer a treatment modality for rape trauma syndrome that is simpler, more rapid, and far less expensive than prior forms of treatment such as exposure therapy. It is also gentler, in the sense that it does not require the person to repeatedly relive the original traumatizing event.

Important though it is to educate people about the reality of rape trauma syndrome, it is equally important to stress that it is not inevitable, for many survivors of rape recover without developing the syndrome. One example of such recovery is that of Elizabeth Smart.[33] As is well known, Smart was abducted at knifepoint from her Salt Lake City home when she was fourteen years old, held captive for nine months, and raped daily, or multiple times a day, over that time span. Yet after being freed, Smart completed high school and college, went abroad on her Mormon mission, testified against her rapist, took up work as a television commentator, started a foundation to advocate for and assist the recovery of survivors of sexual assault, cowrote two books, and is now married with three children—all without any form of professional counseling or therapy. Meanwhile, her rapist is serving a life sentence in federal prison.

CONCLUSIONS

Rape, as well as other forms of sexual assault and harassment, gravely impact the lives of those people—women, for the most part—who experience them. Beyond that, though, they also distort the entire structure of society, casting a veil of fear and mistrust between the sexes and thus hampering cooperation and productivity. Western societies have struggled to eliminate sex crimes, with some success: until recently, at least, those crimes were on a steady decline. Yet they still occur with disturbing frequency. According to RAINN, the largest organization combatting sexual violence in the United States, someone in this country is sexually assaulted every sixty-eight seconds, and yet few of the perpetrators end up in prison.[34]

Rape is a sexual act. Although other factors, such as hostility toward women, may play a role, the primary motivator for rape is a frustrated sex drive. We see that phenomenon across the animal kingdom, from scorpionflies to orangutans, and humans are no different. But whether or not the impulse to rape is translated into action depends on a whole host of factors, including those laid out in the multifactorial models of Beverly McPhail and Neil Malamuth. Perhaps most interesting is Malamuth's demonstration that the capacity for empathy powerfully blocks the expression of sexual violence even when other factors conspire to trigger it.

There may be some individuals—psychopaths—who are congenitally incapable of feeling empathy. The rest of us are born with the capacity for empathy, and it is nurtured during child-hood through a thousand interactions with family, peers, and society at large. Even in adulthood, empathy can be strengthened by specific training programs, according to a meta-analysis of controlled studies.[35] Whether giving sex offenders such training reduces the likelihood of reoffending, however, has not yet been demonstrated.[36] Even if it doesn't reduce reoffending, empa-thy training, in combination with programs to undo false ideas about rape, does seem to increase the willingness of bystanders to intervene in situations involving sexual aggression.[37]

Not all survivors of rape experience the long-lasting ill-effects known as rape trauma syndrome, but for those who do, effective treatments are available. These include exposure therapy and the still-experimental lifespan integration therapy, as well as drugs such as antidepressants and—still on the horizon—psychedelics such as psilocybin.[38]

In the next and final chapter, I turn to a much happier topic: love, the glue that holds most sexual relationships together. Long seen as something mysterious and intangible, a topic for poetry more than for prose, love has in recent years yielded many of its secrets to scientific inquiry.

11

LOVE

Over the course of 25 years, the zoologist Lowell Getz of the University of Illinois at Urbana-Champaign, aided by 49 paid assistants and 1,063 student volunteers and interrupted only by time off for a hernia repair (14 days), rotator cuff surgery (10 days), and cardiac angioplasty (4 days), trapped, marked, and released 236,700 voles in the grasslands of east-central Illinois.[1] From this seemingly mindless activity has blossomed a field of science like no other: a field that seeks to reduce humanity's most highly praised trait—our capacity to establish loving relationships—to a bunch of chemicals.

Voles are small mouselike rodents in the genus *Microtus*. The area that Getz studied is home to two species, the prairie vole (*Microtus ochrogaster*) and the meadow vole (*Microtus pennsylvanicus*). Getz's traps were large enough to accommodate more than one animal, and it wasn't unusual for him to find two or more voles in the same trap. Remarkably, if a trap contained a male and a female prairie vole, there was a good chance that Getz or his assistants would find those same two voles together again weeks or months later, either in the original trap or in another one at a different location.

Intrigued to know whether this was just a coincidence, Getz took some of these male-female pairs and outfitted them with radio collars. He found that they didn't just enter the same traps; they spent nearly all their time together, mostly in their underground burrows. In contrast, if a trap contained a male and female meadow vole, he rarely, if ever, found those two voles together again. Prairie voles form durable pair bonds, in other words, but meadow voles do not.[2]

Male–female pair bonding is not very common in the animal world. Birds are the big exception: many avian species form breeding pairs, either for a single breeding season or occasionally for life. This behavioral pattern allows for the division of labor between foraging and the incubation of eggs or the protection of nestlings. Among mammals, the usual pattern is for males and females to get together only for mating, then they go their separate ways. Some mammals form stable breeding groups consisting of a single male with several females—a "harem." Only about 9 percent of mammalian species form pair bonds, but those that do are scattered widely across the class. They include North American beavers, gibbons, titi monkeys, jackals, humans, otters—and prairie voles.[3]

The existence of a pair bond doesn't necessarily imply sexual exclusivity, a fact that has been revealed by DNA paternity testing. In the case of prairie voles, a pair-bonded female will sometimes mate with a "wanderer," a solitary male who just happens to be in the vicinity, like the cable repair guy in 1990s pornographic videos.[4] Nor is the pair bond unbreakable: Getz and his colleagues found that putting a pair-bonded female with a new male for just eight days caused her to bond with the new male, after which she would treat her original partner as an unwelcome stranger. Still, for as long as the bonded couple stays together, they are truly an item, sharing household chores, coparenting, and, if nothing else is on the agenda, just cuddling. Such behavior is completely foreign to their close relative, the meadow vole, or to several other species of nonmonogamous voles.

Pair-bonded prairie voles act as if they love each other, but that love has a dark side. It's counterbalanced by a marked increase in hostility toward other voles, especially other voles of the same sex. A vole that previously might have treated a stranger with curiosity or, at worst, indifference will now unleash ferocious attacks until the interloper gets the message and beats a retreat. Thus whatever process triggers bonding also triggers aggressiveness toward outsiders. Presumably this aggressiveness serves to maintain the exclusivity of the pair bond.

LOVE IN THE LAB

One of Getz's long-time assistants was Sue Carter, who is now an emerita professor of biology at the University of Indiana. Carter brought vole research indoors; she showed that voles could be raised and bred in the lab and developed a three-chamber setup to study the animals' partner preferences. If a virgin male and female prairie vole are kept together but not allowed to mate, they form a pair bond over a period of a few days. But if they are allowed to mate, which they do repeatedly in a single day, they form a bond much more rapidly: the female bonds within about six hours, but the male takes a bit longer—about twenty-four hours.

So something happens in the brains of voles when they mate that focuses their sexual and social interest on each other. Carter and her colleagues showed that two chemicals produced by neurons in the hypothalamus, OXYTOCIN and VASOPRESSIN, play key roles in the formation of pair bonds between male and female prairie voles. These chemicals are PEPTIDES, which are chains of amino acids, like proteins, but much shorter. Oxytocin and vasopressin consist of just nine amino acids, which form a ring with a tail, like the letter Q. Seven of the nine amino acids are identical in the two peptides, and oxytocin and vasopressin probably evolved from a single peptide in the very distant past.

Oxytocin and vasopressin are hormones that are transported from the hypothalamus—much of it from a cell group called the PARAVENTRICULAR NUCLEUS—to the pituitary gland, where they are secreted into the bloodstream (see figure 11.1). Their hormonal roles have been known for a long time: oxytocin causes contraction of the muscular coat of the uterus and thus enables the delivery of young (childbirth in humans), and it also triggers milk let-down in response to suckling at the breast. Vasopressin acts on the kidney to retain fluid and causes contraction of

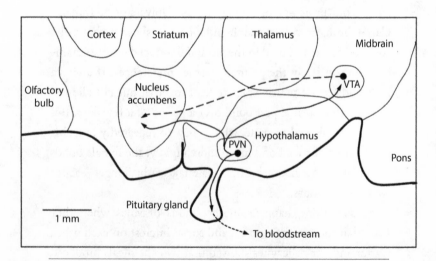

FIGURE 11.1. Oxytocin pathways.

Source: Figure by the author.

small arteries; both actions tend to raise blood pressure. More recently, though, it was discovered that oxytocin and vasopressin are also released within the brain, and several parts of the brain are rich in receptors for one or the other peptide or for both. In the context of their activity within the brain, the two peptides are referred to as NEUROMODULATORS rather than hormones.

Carter's team found that injecting oxytocin directly into the brains of female prairie voles mimicked the effects of mating; that is, the vole would rapidly form a pair bond with a male even without mating. Conversely, injecting a drug that blocked the effects of oxytocin would prevent the formation of a pair bond even if the female did mate with a male.

Vasopressin has similar actions, and in fact there is crosstalk between the two peptides, because oxytocin receptors are somewhat sensitive to vasopressin and vasopressin receptors to oxytocin. Nevertheless, vasopressin appears to play a more important role in males than in females. In any event, oxytocin and vasopressin together constitute key players in the formation of pair bonds in prairie voles.

Vole species that don't form pair bonds—meadow voles or another nonmonogamous species, montane voles—cannot be induced to do so by injection of oxytocin or vasopressin into their brains. So it's not just the presence or absence of these peptides in the brain that determine whether the animal forms pair bonds. Rather, there is something about the way the brain responds to the peptides that differs between species.

Starting in the 1990s, several researchers began tackling the pair-bonding phenomenon with molecular genetic techniques. These researchers included Thomas Insel at the National Institute of Mental Health, Zuo-Xin Wang at the University of Massachusetts, and Larry Young at Emory National Primate Research Center. They showed that a receptor for vasopressin was distributed differently in the brains of prairie and montane voles.[5] (There are three vasopressin receptors; the receptor involved in pair-bonding is named "V1a.") Of particular interest, a structure named the VENTRAL PALLIDUM was rich in these vasopressin receptors in prairie voles but almost devoid of them in montane voles. The ventral pallidum sits close to and receives outputs from the nucleus accumbens, whose role in sexual behavior I discussed in chapter 3. Like the nucleus accumbens, it is part of the brain's reward system, and dopamine is released there during rewarding behaviors. Blocking vasopressin receptors in just the ventral pallidum was sufficient to prevent pair bonding in prairie voles; conversely, adding vasopressin receptors to the ventral pallidum of meadow voles caused them to form pair bonds, something that regular meadow voles never do.[6]

Drilling down even further into the biology, Young, Insel, and their colleagues sought to determine if there were differences between the gene for the vasopressin receptor in monogamous and nonmonogamous species. Indeed, there were—not in the part of the gene that actually codes for the receptor but in a neighboring portion of the gene that contains elements controlling where and when the gene is active. Here Young's group found a stretch of what is called MICROSATELLITE DNA, where the DNA, as if caught in a stutter, repeats a short sequence of DNA bases over and over. This DNA was present in prairie voles as well as in another species that forms pair bonds, the pine vole, but was missing in two species that don't form pair bonds, meadow and montane voles.[7]

Was the presence of this piece of DNA the key that allowed for pair bonding? To test this idea, Young created a strain of TRANSGENIC mice possessing the prairie vole's gene for the vasopressin receptor, including the microsatellite DNA. In these mice, the distribution of vasopressin receptors in the brain was unlike that seen in regular mice and instead resembled that seen in prairie voles. What's more, when a male mouse with the prairie vole DNA was placed with a normal female mouse, it was far more intimate with the female than usual, staying close to her and grooming her significantly more than a regular male mouse would do. That behavior may not have amounted to pair bonding, but it certainly was an indication of increased sociability directed toward a sex partner.

Because pair bonding is so durable in prairie voles—lifelong, in most cases—a group at Florida State University has investigated whether there are changes to the voles' genome at the time the pair bond is initiated.[8] Indeed there are: when voles mate for the first time, the genes for both the oxytocin and vasopressin receptors undergo so-called epigenetic modification. As I briefly mentioned in the context of sexual orientation (chapter 4), this process involves changes to chromosomes that do not affect the actual DNA sequence. Rather, in the case of the FSU study, the changes were to HISTONES—proteins that provide structural support to DNA and control its activity. Such changes are known to be involved in the formation of lasting memories. When voles mate, the epigenetic modification of histones causes the density of oxytocin and vasopressin receptors in the nucleus accumbens to increase and remain high. The precise sequence of changes is different between females and males, even though the behavioral result—pair bonding—is the same.

What about those wanderers—the cable repair voles, as I like to call them? What causes them to invade the territories of pair-bonded voles in search of promiscuous sex, and why don't they form pair bonds of their own? At least part of the reason may be genetic. Young's group found a region near the gene for the prairie voles' oxytocin receptor in which the DNA sequence varies slightly between individuals.[9] Prairie voles with one version (allele) of this DNA sequence have fewer oxytocin receptors in the nucleus accumbens than voles with the other allele, and they don't show any partner preferences in the kind of laboratory test devised by Carter. They are likely to be the wanderers.

It's not clear which behavior pattern—pair bonding or promiscuity—is the better strategy in terms of producing the most offspring. Whichever pattern is better, one might expect the allele promoting that behavior to outcompete and eliminate the other over the course of the generations, so that eventually all prairie voles would show the more successful behavior. But it's not that simple, because the two reproductive strategies feed off each other. If pair bonding is more successful, for example, the numbers of wanderers will go down, but that reduces competition among wanderers and improves their chances of scoring with pair-bonded individuals. Therefore an equilibrium situation is reached in which the two alleles, and the two behaviors, are present in stable proportions. In one study, about 75 percent of prairie voles were pair bonders and 25 percent were wanderers. (Some females are wanderers, too.)

Environmental conditions can affect what proportion of prairie voles are pair bonders or wanderers. If food becomes scarce, for example, the overall density of the vole population will go down, so a wanderer has to travel farther to find sex partners and will find fewer of them in the course of a day's wandering. Wandering thus becomes less reproductively efficient, and that will favor the pair-bonding strategy. Because environmental conditions vary from place to place, the proportions of pair bonders and wanderers differ at different locations, as well as at different times in the same location. The existence of alleles favoring two different reproductive strategies therefore confers flexibility on the species as a whole and favors its survival.

WHAT PSYCHOLOGISTS TELL US

Before getting into what we know about the biology of love and pair bonding in humans, I should mention some key ideas about the psychology of love in our species. This field was pioneered by Ellen Berscheid and Elaine Hatfield of the University of Minnesota (Hatfield is now at the University of Hawaii). In 1969, they proposed that romantic love has two elements: PASSIONATE LOVE and COMPANIONATE LOVE. Robert Sternberg of Cornell University later divided companionate love into two components: INTIMACY and COMMITMENT. This then led to Sternberg's well-known TRIANGULAR THEORY of love, whereby love between sex partners has three possible components—passion, intimacy, and commitment—which may be combined in various proportions. Romantic love, for example, is the combination of passion and intimacy.

Passion is being head-over-heels in love. According to Berscheid and Hatfield, it's a yearning for union with another person. If that union is achieved, passion tends to fade over time. You can't yearn for something you already have, after all. (Some psychologists believe that certain couples maintain passion indefinitely or, more plausibly, that passion can be reignited by life events such as temporary separation.)

Passion is triggered by sexual attraction, by "chemistry"— the sense that the two of you fit together as soulmates—and by reciprocity—the belief that the other person has developed an interest in you. Of course, that belief may be mistaken, so passionate love can be entirely one-sided, as most of us have experienced at some point in our lives, either as the rejector or the rejectee. And if passion dies without being replaced by intimacy or commitment, the lover may look back with embarrassment or incredulity at their infatuation. "Methought I was enamour'd of an ass," lamented Shakespeare's Titania. "O, how mine eyes do loathe his visage now."

Intimacy is what it sounds like: a strong liking based on self-disclosure and deep familiarity between partners. For many couples, physical and sexual contacts reinforce intimacy and help maintain it over a lifetime, even if sex loses some of its novelty and excitement over the years.

Commitment is the cognitive component of love—the decision to love and keep loving despite any difficulties that may arise. Commitment is at the heart of traditional Christian marriage vows: "I promise to love, honor," and so on.

But surely one can't promise to love someone, any more than one can promise to love broccoli? Isn't love something that just happens to you? Well, maybe not. The humanistic social psychologist Erich Fromm, who died in 1980, believed strongly in the cognitive aspects of love. "One neglects to see an important factor in erotic love, that of *will*," he wrote. "To love somebody is not just a strong feeling—it is a decision, it is a judgment, it is a promise."[10] The importance of commitment as a building block of love is borne out by traditional arranged marriages, in which a couple who barely knew each other at the time of their wedding may nevertheless develop a bond as deep as that of any couple brought together by passion.

The distinction between the first two components of love—passion and intimacy—is widely understood and acknowledged across cultures. This is how a woman of the hunter-gatherer !Kung people of southern Africa explained the matter to anthropologist Marjorie Shostak: "When two people come together their hearts are on fire and their passion is very great. After a while, the fire cools and that's how it stays." An established married relationship, on the other hand, is "rich, warm and secure."[11] Regarding the third component, commitment, people know what it is, but they don't always see it as being a component of love.

Sternberg's original study, published in 1986, was based on multiple-item questionnaires filled out by undergraduate students, and the three components were derived by factor analysis—a statistical procedure that looks for groupings within a large data set. Some later researchers failed to confirm the three factors, especially when studying different populations. Nevertheless, a 2021 study with 114 authors (including Sternberg) confirmed the original findings in a diverse sample of 7,332 individuals in 25 countries. It further confirmed that passion tends to decrease over the course of a relationship, whereas commitment increases.[12]

So which, if any, component of human love has some commonality with pair bonding in prairie voles? Hardly passion, except possibly for the briefest of intervals when the voles first get it on. And hardly commitment, either: prairie vole couples behave as if they are committed to each other, certainly, but a cognitive basis for that behavior is too much to ask of a vole—a creature whose intellectual capacity is about that of a wind-up toy. It's the intimacy of voles' pair bonds that offers a possible parallel or model for human love—for the intimacy component of Sternberg's triangle.

Intimacy is also evident in those few nonhuman primate species that form pair bonds. A well-known example is that of titi monkeys, a group of New World monkey species, all of which demonstrate affection between pair-bonded couples. A bonded pair likes to sit in close contact on a branch, with their entwined tails hanging beneath. They also "duet," or interlace their calls, and they share parental duties.

One significant difference between the pair bonds of prairie voles and titi monkeys, however, lies in the senses the two species use to identify their mates. Prairie voles rely primarily on olfaction, whereas titi monkeys are highly visual animals and probably recognize their mates by sight. Consistent with this difference, a group at the University of California, Davis, where a breeding colony of titi monkeys is housed, found that the distribution of vasopressin receptors in the brains of titi monkeys is significantly different from that seen in prairie voles. In the voles, they are densely distributed within olfactory brain structures and not in the visual system, whereas in titi monkeys, it's the other way around.[13]

OXYTOCIN IN HUMANS

Differences of the kind just described should make us cautious about extrapolating from prairie voles to humans. We are only distantly related to voles, after all: our most recent common ancestors lived about 75 million years ago. Nevertheless, many researchers have investigated the idea that oxytocin and/or vasopressin play roles in interpersonal bonding in humans as they do in prairie voles.

The first evidence for such a role came from work on mother–child attachment. That form of attachment probably evolved first, and it later provided the "wetware" for other forms of attachment, including father–child attachment, social living, and sexual pair bonding. With current technology, it's not possible to monitor the changing levels of oxytocin or vasopressin within the living brain. Researchers therefore measure these levels in the blood, which are dependent on secretion of the peptides from the pituitary gland. The main findings are that maternal oxytocin levels rise during mother–infant interactions and that steady-state oxytocin levels are predictive of the quality of mother–infant interactions over time.[14]

Because mating and the accompanying rise in oxytocin and vasopressin levels play such an important role in the initiation of pair bonds in prairie voles, it's of interest to know whether anything comparable happens in our own species. There's no automatic connection between sex and pair bonding in humans of course. For example, most one-night stands are just that. Yet they can also be something more. According to Helen Fisher, a biological anthropologist at Indiana University's Kinsey Institute who is well known for her research and writings on the topic of romantic love, more than one in four Americans have experienced a one-night stand that turned into a long-term relationship.[15] So sexual intercourse, even in a casual context, may sometimes trigger the formation of pair bonds à la prairie vole.

It turns out that vasopressin levels remain more or less unchanged during human sexual activity. The findings on oxytocin are more interesting. In both sexes, there's a surge of oxytocin at the time of orgasm, followed by a return to baseline levels within a few minutes or tens of minutes.[16] The oxytocin surge strengthens the contractions in the genital tract that accompany orgasm. If oxytocin levels are artificially raised by the use of a nasal spray, the subjective sensation of orgasm is more intense.[17] Conversely, if the secretion of oxytocin is blocked, the sensation of orgasm is blunted.[18]

Aside from the brief surges of oxytocin that occur at orgasm, the steady baseline levels of oxytocin are related to a person's partnership status. In an Israeli study from the laboratory of Ruth Feldman, with Inna Schneiderman as first author, the oxytocin levels of new lovers were twice as high as those of unattached individuals, and those levels remained high for at least six months.[19] The new lovers who had the most intimate interactions had the highest oxytocin levels. The researchers even found that the future of a romantic relationship was somewhat

predictable based on oxytocin levels soon after its commence-
ment: men and women who had split up by the time of the six-
month assessment had significantly lower oxytocin levels at the
time of the initial assessment than those who stayed together.

Numerous studies have tested the effects of nasally administered oxytocin on perceptions of attractiveness, trust, and the
like. There has been an ongoing dispute about whether oxytocin sprayed into the nose actually gets into the brain, but
accumulating evidence indicates that it does so, probably by
passing through nerve channels in the roof of the nasal cavities.[20]
Although transport of oxytocin via this route is probably inef-
ficient, researchers administer such huge doses—more than is
present in the body's entire natural storehouse of oxytocin, the
pituitary gland—that enough enters the brain to exert measurable psychological effects.

An early study reported that nasal oxytocin increased trust in
the context of a money game, which involved giving money to
another player with the expectation of reciprocity. But a recent
attempt to replicate that study failed to do so, which means that
the much-publicized connection between oxytocin and trust sits
in a kind of scientific limbo pending further investigation.[21]

Regarding attractiveness, a German research group, with the
psychologist Dirk Scheele as first author, examined the effects of
intranasal oxytocin on men in heterosexual partnerships.[22] The
men had been partnered for about three years but described themselves as still passionately in love with their partners. They were
asked to rate the attractiveness of photos of their partners and
of unfamiliar women. Oxytocin increased the perceived attractiveness of their partners but not that of the unfamiliar women,
whereas a placebo (inactive spray) had no effect in either case.

The researchers also scanned the men's brains while they
viewed the photos. When the men viewed their partners,

oxytocin strongly increased activity in the nucleus accumbens compared with a placebo, but it did not affect the activity when the men viewed unfamiliar women. These and other experiments suggest that oxytocin may be involved in the maintenance of human pair bonds; whether it is involved in the initial formation of pair bonds, as is the case for voles, is not yet clear. Like voles, humans have oxytocin receptors in the nucleus accumbens, so it's likely that the nasally administered oxytocin had a direct effect on the nucleus.

Interestingly, oxytocin and dopamine receptors in the nucleus accumbens are yoked together in pairs or clusters on the surface of neurons.[23] This arrangement likely ensures that the neurons are activated most strongly when dopamine and oxytocin are released simultaneously—from terminals of axons that come from the ventral tegmental area and from the hypothalamus, respectively. Speculatively, a coincidence-detecting mechanism of this kind could underlie important aspects of pair bonding, such as the role of orgasmic pleasure in the company of a specific partner.

As with prairie voles, individual humans vary in how likely they are to form pair bonds and remain in them. There are also slight differences between individuals in the genes that code for the oxytocin and vasopressin receptors. (As with voles, I'm referring to the vasopressin receptor known as V1a.) Thus it's of interest to know whether different alleles of either of these genes influence pair bonding, as is true for voles.

In 2008, a group led by Hasse Walum of Stockholm's Karolinska Institute reported on a repeating DNA sequence (microsatellite) known as RS3, which is located close to the gene for the human vasopressin receptor.[24] The number of DNA repeats at this location is variable. Walum's group reported that one allele, which is characterized by a relatively large number of repeats, was more common in married than unmarried men. Among married men, this allele was associated with stronger spousal bonding as assessed by both the men and their wives. Among women, on the other hand, the researchers detected no association between RS3 alleles and relationship quality. This finding may correspond with findings in prairie voles, some of which indicate that vasopressin is more involved in pair bonding in males than in females.

Regarding the oxytocin receptor gene, attention has focused on a specific base in the gene that can be either A (adenine) or G (guanine). This is a single nucleotide polymorphism, or SNP (as described in chapter 4), and is known as RS53576. Because chromosomes come in homologous pairs, a person can have two A alleles, one A and one G, or two Gs (AA, AG, or GG). Several studies have reported that people who possess the GG combination are more empathetic, altruistic, and satisfied with sexual activity within a pair bond compared with AA individuals, while AG individuals are intermediate in those traits.[25]

Yet another genetic variation relevant to pair bonding involves a gene named CD38. The protein product of this gene regulates the release of oxytocin within the brain. Within the gene there is an SNP named RS3796863, at which the DNA base can be either A or C (cytosine). In a study from the laboratory of Jennifer Bartz at McGill University, couples in which one partner possessed CC (cytosine on both chromosomes) engaged in more intimate behavior than couples in which neither partner had that combination.[26] Intriguingly, each partner's intimate behavior was more frequent regardless of whether it was that person or their partner who possessed the CC pattern. This finding highlights how partners influence each other in the development of intimacy. Couples in which both partners possessed CC scored higher than any other couples, however. It's not yet known how individuals with the AA, AC, and CC alleles differ in terms of where and how much oxytocin is released in the brain.

All these findings regarding genetic differences that affect the quality of romantic relationships must be treated with some caution, because the numbers of persons studied and the number of studies on each topic so far have been low. The field of behavioral genetics as a whole has been plagued by contradictions between studies, and the effects of individual genes on psychological traits typically have been small. It's too early for people to have their prospective partners checked for the magic CC combination of rs3796863 alleles, for example. Still, at least some of the findings will turn out to be robust, and eventually we'll know enough about the genetics of romance to offer biologically based advice to prospective couples—if anyone should be so rash as to trust science over their own gut instincts.

LOVE IN THE SCANNER

The first functional imaging study of romantic love was carried out by Andreas Bartels and Semir Zeki in Zeki's laboratory at University College London.[27] Like myself, Zeki turned to the study of sexuality after a long and—in his case—distinguished career focused on the visual system. Bartels and Zeki advertised for volunteers who were "truly, deeply, and madly in love." That sounds like undiluted passion, but the actual participants had been in their relationships for a considerable length of time (a median of 2.3 years), so it's likely that passion had been supplemented or even replaced by intimacy in many cases.

Bartels and Zeki compared the effects on brain activity of viewing a photo of a participant's beloved with the effects of viewing photos of nonsexual friends. They found increased activity in the anterior cingulate cortex, an area that's involved in sexual arousal, as I described in chapter 3. Conversely, they found decreased activity in a cortical region farther back, called the POSTERIOR CINGULATE CORTEX, that appears to be involved in some negative emotions such as pain, depression, and anxiety.

On the face of it, these changes seem very appropriate. Being reminded of the beloved increases positive feelings and damps down negative ones. Whether it's quite that simple is doubtful, however, because both the anterior and posterior cingulate cortices comprise several subregions that are thought to serve distinct functions.

Bartels and Zeki, as well as other investigators who have carried out similar studies, also found activations of several subcortical regions known to be involved in emotional processing. One way to summarize these studies is to say that the activity patterns resemble those seen in a variety of rewarding scenarios, such as receiving a euphoric drug or a monetary prize. Many of the active regions are rich in dopamine and oxytocin receptors. One particularly active region is a small zone in the brainstem known as the ventral tegmental area. That area contains the neurons whose axons travel to the nucleus accumbens and other forebrain regions and release dopamine there.

It's clear, then, that seeing and thinking about a person with whom one is happily in love activates reward circuitry in the brain, as one might expect. But is there anything different between this "romantic reward" and other kinds of reward? One observation relevant to this question was made in 2020 by Bianca Acevedo of the University of California, Santa Barbara, and three colleagues.[28] When viewing photos of their beloveds, Acevedo's volunteers showed activation not only of reward circuitry but also of the region of primary sensory cortex that carries the representation of the genitals. That happened even though there was no actual stimulation of the genitals—the usual sensory experience that causes activity in that region—nor were the volunteers even thinking about sex during the scans. Acevedo speculated that some kind of conditioning process happens over the course of a sexually intimate romantic relationship, such that the genital sensory region becomes linked into the reward circuitry activated by seeing the beloved. Regardless of the exact mechanism, this finding is consistent with the obvious fact that romantic love differs from other rewarding experiences by virtue of its erotic nature.

Most of the brain imaging studies that have been done on romantic love have focused on established, happy relationships—even the newlyweds studied by Acevedo had been living together for a couple of years before getting married. No one has really studied the process of falling in love or the state of having fallen in love before or without reciprocation by the loved person (unrequited love). In other words, the brain basis of pure passionate love—Berscheid and Hatfield's "yearning for union with another person"—hasn't yet been identified.

I may be a hopeless unromantic, but it seems to me that negative emotions—craving, anxiety, sadness, despair, and jealousy—are just as central to the experience of passionate love as the joy of having that love reciprocated. Two studies have attempted to address those facets of love by examining brain activity patterns in women or men who had been jilted. The participants had recently experienced the breakup of an established romantic relationship, but they were still passionately in love and yearned to be reunited with their rejector. They expressed thoughts like "I think about him constantly"; "It hurts so much. I crumble. I just start crying"; "What's the point without her?"; and "How I miss his smile and his eyes."

In one of the studies, led by Helen Fisher of Rutgers University, the participants were tested in a paradigm similar to that used in the studies described previously—that is, they were scanned while viewing a photo of their rejector or a neutral acquaintance.[29] Brain regions that were specifically active during viewing of the rejector included many of the same areas as in "happily in love" persons: the ventral tegmental area, nucleus accumbens, ventral pallidum, and anterior cingulate cortex. This is perhaps not surprising because the participants were being rewarded during the scan—with the image of their otherwise inaccessible beloved.

In the other study, led by the psychiatrist Arif Najib of the University of Tübingen, Germany, the participants—all women—did not view any photographs. Rather, they were instructed simply to engage in the kind of sad thoughts about their ex that currently dominated their mental lives: "My ex telling me it's not working out and it's over"; "Our mutual dreams—not a shared dream anymore"; and the like. In alternating runs, the participants had emotionally neutral thoughts about an acquaintance.[30]

The findings were very different from those of Fisher's group. Several of the reward-linked regions—the nucleus accumbens, ventral pallidum, brainstem, and anterior cingulate, as well as a broad expanse of the prefrontal cortex—were actually deactivated; that is, they were less active when thinking about the ex than when thinking about the acquaintance. Only a few regions were more active when thinking about the ex, but these included the posterior cingulate cortex. That zone, as mentioned earlier, is associated with negative emotions and was found to be deactivated in Bartels's and Zeki's study of happily-in-love individuals. In other words, happy and unhappy lovers experience patterns of brain activity that, to a considerable degree, mirror each other, yet both kinds of lovers are authentically in love.

These findings suggest that there's something more basic to love, something independent of the emotions it engenders, that remains to be captured by neuroscientists. This more basic element may be the mechanism that links the happy or unhappy emotions of love to one specific individual: the beloved.

It's known that the identity of individuals can be encoded in the activity of single neurons or clusters of neurons. In one well-known study, for example, a neuron in a person's hippocampus responded to photos or drawings of the actor Halle Berry or to her printed name but not to images or names of other individuals or objects that were tested.[31] We could speculate that the neural representation of a person with whom one falls in love expands and becomes more strongly connected to the emotion-mediating pathways already described.

It will take major technical advancements, however, to demonstrate such a phenomenon. For one thing, single-cell recordings deep in the brain can be done only in the context of neurosurgical procedures for conditions such as epilepsy, and usually only for a few minutes. In addition, these person-specific neurons appear to be intermingled with other neurons in a way that would make it difficult to visualize their activity with current scanning technologies.

CONCLUSIONS

Though incomplete, biological studies of love have been revelatory. We know that there are marked parallels between what goes on in lovers' brains and what binds two prairie voles together. In both cases, the bond is forged and maintained by the brain's central reward system, involving the activation of neurons in the nucleus accumbens and other structures by dopamine and the facilitation of that activation by oxytocin and vasopressin. In both cases, slight differences between individuals in the genes for these peptides can have large effects on an individual's capacity to form and maintain loving bonds.

In all these respects, we are little different from voles. What is distinctly human, though, is our capacity to think about our actual or prospective relationships, give names to our feelings, and interpret our actions—sometimes in inaccurate ways. Sue Carter, a prime mover in the oxytocin hypothesis of love, put it this way in an essay coauthored with Stephen Porges: "The biology of love originates in the primitive parts of the brain— the emotional core of the human nervous system—that evolved long before the cerebral cortex. . . . The modern cortex struggles to interpret the primal messages of love, and weaves a narrative around incoming visceral experiences, potentially reacting to that narrative rather than reality."[32]

Looking at research in the entire field of human sexuality, I'm struck by how effectively it has brought together researchers from many disciplines—sociology, psychology, genetics, and neuroscience in particular. These interactions are bringing clarity to a topic that is central to our lives yet enmeshed in prejudice and social conflict.

As with science in general, findings in the area of sexuality are tentative; studies contradict each other; theories become outmoded. If at some point I get to revise this book for a second edition, I'm sure I'll find statements and ideas that I'll regret having included. But also, there will be new findings and new ideas that will build on and strengthen the shaky foundations that have been constructed so far.

NOTES

1. WHY HAVE SEX?

1. C. M. Meston, and D. M. Buss, "Why Humans Have Sex," *Archives of Sexual Behavior* 36 (2007): 477–507.
2. B. W. Hewlett and B. L. Hewlett, "Sex and Searching for Children Among Aka Foragers and Ngandu Farmers of Central Africa," *African Study Monographs* 31 2010): 107–25.
3. A. Sonfield, K. Hasstedt, and R. Benson Gold, *Moving Forward: Family Planning in the Era of Health Reform* (report, Guttmacher Institute, Washington, DC, March 2014), https://www.guttmacher.org/report /moving-forward-family-planning-era-health-reform.
4. Technical terms that appear more than once are listed in the glossary.
5. J. Maynard Smith, *The Evolution of Sex* (Cambridge: Cambridge University Press, 1978).
6. A. K. Gibson, L. F. Delph, and C. M. Lively, "The Two-Fold Cost of Sex: Experimental Evidence from a Natural System," *Evolution Letters* 1 (2017): 6–15.
7. L. H. Liow L. Van Valen, and N. C. Stenseth, "Red Queen: From Populations to Taxa and Communities," *Trends in Ecology and Evolution* 26 (2011): 349–58; M. Ridley, *The Red Queen: Sex and the Evolution of Human Nature* (New York: Harper Perennial, 2003).
8. A. K. Gibson et al., "Periodic, Parasite-Mediated Selection for and Against Sex," *American Naturalist* 192 (2018): 537–51.

9. A. Kong et al., "Rate of de Novo Mutations and the Importance of Father's Age to Disease Risk," *Nature* 488 (2012): 471–75.

10. J. R. Peck, "A Ruby in the Rubbish: Beneficial Mutations, Deleterious Mutations and the Evolution of Sex, *Genetics* 137 (1994): 597–606.

11. M. J. McDonald, D. P. Rice, and M. M. Desai, "Sex Speeds Adaptation by Altering the Dynamics of Molecular Evolution," *Nature* 531 (2016): 233–36.

12. T. Kuroda-Kawaguchi et al., "The AZFc Region of the Y Chromosome Features Massive Palindromes and Uniform Recurrent Deletions in Infertile Men," *Nature Genetics* 29 (2001): 279–86; S. Rozen et al., "Abundant Gene Conversion Between Arms of Palindromes in Human and Ape Y Chromosomes," *Nature* 423 (2003): 873–76.

13. D. Crews, M. Grassman, and J. Lindzey, "Behavioral Facilitation of Reproduction in Sexual and Unisexual Whiptail Lizards," *PNAS* 83 (1986): 9547–50.

14. A. A. Lutes et al., "Laboratory Synthesis of an Independently Reproducing Vertebrate Species," *PNAS* 108 (2011): 9910–15.

15. J. Gutekunst et al., "Clonal Genome Evolution and Rapid Invasive Spread of the Marbled Crayfish," *Nature Ecology & Evolution* 2 (2018): 567–73.

16. B. Jenni, "Bdelloid Rotifers: So Common Yet So Weird!", https://www.youtube.com/watch?v=rIUxqKR-KrM.

17. D. B. Mark Welch and M. Meselson, "Evidence for the Evolution of Bdelloid Rotifers Without Sexual Reproduction or Genetic Exchange," *Science* 288 (2000): 1211–15.

18. E. A. Gladyshev, M. Meselson, and I. R. Arkhipova, "Massive Horizontal Gene Transfer in Bdelloid Rotifers," *Science* 320 (2008): 1210–13; N. Debortoli et al., "Genetic Exchange Among Bdelloid Rotifers Is More Likely Due to Horizontal Gene Transfer Than to Meiotic Sex," *Current Biology* 26 (2016): 723–32.

19. B. Ponting, "SoCal Woman, 45, Unexpectedly Gives Birth, Didn't Know She Was Pregnant," https://tinyurl.com/ppdspvkhFox 5, October 26, 2017, https://fox5sandiego.com/news/california-woman-45-unexpectedly-gives-birth-after-not-realizing-she-was-pregnant/.

20. F. Bruni, "Woman, 29, Still in 10-Year Coma, Is Pregnant by a Rapist," *New York Times*, January 25, 1996, https://www.nytimes.com/1996/01/25

/nyregion/woman-29-still-in-10-year-coma-is-pregnant-by-a-rapist
.html; L. Yglesias, "Sad and Silent B'day for Mom in Coma," *New York Daily News*, April 28, 1996.

21. J. G. Milhaven, "Thomas Aquinas on Sexual Pleasure," *Journal of Religious Ethics* 5 (1977): 157–81.

22. F. B. M. de Waal, "Bonobo Sex and Society," *Scientific American* 272 (March 1995): 82–88.

2. ATTRACTION

1. V. Swami and M. J. Tovee, "Does Hunger Influence Judgments of Female Physical Attractiveness?" *British Journal of Psychology* 97 (2006): 353–63.

2. L. D. Nelson and E. L. Morrison, "The Symptoms of Resource Scarcity: Judgments of Food and Finances Influence Preferences for Potential Partners," *Psychological Science* 16 (2005): 167–73.

3. G. M. Adahada, "Fattening, a Traditional Institution in Annang Antiquity: Its Leadership Role and Limitations," *Journal of African Studies and Sustainable Development* 3 (2020): 74–86.

4. M. J. Tovee et al., "Changing Perceptions of Attractiveness as Observers Are Exposed to a Different Culture," *Evolution and Human Behavior* 27 (2006): 443–56.

5. L. G. Boothroyd et al., "Television Exposure Predicts Body Size Ideals in Rural Nicaragua," *British Journal of Psychology* 107 (2016): 752–67.

6. A. E. Becker, "Television, Disordered Eating, and Young Women in Fiji: Negotiating Body Image and Identity During Rapid Social Change," *Culture, Medicine and Psychiatry* 28 (2004): 533–59.

7. W. D. Lassek and S. J. C. Gaulin, "Do the Low WHRs and BMIs Judged Most Attractive Indicate Better Health?" *Evolutionary Psychology* 16 (2018), https://doi.org/10.1177/1474704918803998; Lassek and Gaulin, "Do the Low WHRs and BMIs Judged Most Attractive Indicate Higher Fertility?" *Evolutionary Psychology* 16 (2018), https://doi.org/10.1177/1474704918800063.

8. A. Sell, A. W. Lukazsweski, and M. Townsley, "Cues of Upper Body Strength Account for Most of the Variance in Men's Bodily Attractiveness," *Proceedings of the Royal Society of London. Series B: Biological Sciences* 284 (2017), https://doi.org/10.1098/rspb.2017.1819.

9. L. H. Lidborg, C. P. Cross, and L. G. Boothroyd, "A Meta-Analysis of the Association Between Male Dimorphism and Fitness Outcomes in Humans," *Elife* 11 (2022), https://doi.org/10.7554/eLife.65031.

10. G. Korte, "With Women in Combat Roles, a Federal Court Rules Male-Only Draft Unconstitutional," *USA Today*, February 24, 2019.

11. M. Andersson and Y. Iwasa, "Sexual Selection," *Trends in Ecology and Evolution* 11 (1996): 53–58.

12. B. C. Jones et al., "No Compelling Evidence That More Physically Attractive Young Adult Women Have Higher Estradiol or Progesterone," *Psychoneuroendocrinology* 98 (2018): 1–5.

13. V. S. Johnston, R. Hagel, M. Franklin, B. Fink and K. Grammar, "Male Facial Attractiveness: Evidence for Hormone Mediated Adaptive Design," *Evolution and Human Behavior* 22 (2001): 251–67.

14. M. H. Larmuseau, K. Matthijs, and T. Wenseleers, "Cuckolded Fathers Rare in Human Populations," *Trends in Ecology and Evolution* 31 (2016): 327–29.

15. B. C. Jones et al., "General Sexual Desire, But Not Desire for Uncommitted Sexual Relationships, Tracks Changes in Women's Hormonal Status," *Psychoneuroendocrinology* 88 (2018): 153–57.

16. Z. Yang and J. C. Schank, "Women Do Not Synchronize Their Menstrual Cycles," *Human Nature* 17 (2006): 433–47.

17. G. Rhodes et al., "Attractiveness of Own-Race, Other-Race, and Mixed-Race Faces," *Perception* 34 (2005): 319–40.

18. E. V. Stepanova and M. J. Strube, "Attractiveness as a Function of Skin Tone and Facial Features: Evidence from Categorization Studies," *Journal of General Psychology* 145 (2018): 1–20.

19. M. B. Lewis, "Why Are Mixed-Race People Perceived as More Attractive?", *Perception* 39 (2010): 136–38.

20. M. Babel, G. McGuire, and J. King, "Towards a More Nuanced View of Vocal Attractiveness," *PLoS One* 9 (2014): e88616, https://doi.org/10.1371/journal.pone.0088616.

21. C. Wedekind et al., "MHC-Dependent Mate Preferences in Humans," *Proceedings of the Royal Society of London. Series B: Biological Sciences* 260 (1995): 245–49.

22. C. Wedekind and S. Füri, "Body Odour Preferences in Men and Women: Do They Aim for Specific MHC Combinations or Simply

Heterozygosity?", *Proceedings of the Royal Society of London. Series B: Biological Sciences* 264 (1997): 1471–79.

23. F. Probst et al., "Men's Preferences for Women's Body Odours Are Not Associated with Human Leucocyte Antigen," *Proceedings of the Royal Society of London. Series B: Biological Sciences* 284 (2017), https://doi.org /10.1098/rspb.2017.1830.

24. T. Kimchi, J. Xu, and C. Dulac, "A Functional Circuit Underlying Male Sexual Behaviour in the Female Mouse Brain," *Nature* 448 (2007): 1009–14.

25. M. J. Baum and J. A. Cherry, "Processing by the Main Olfactory System of Chemosignals That Facilitate Mammalian Reproduction," *Hormones and Behavior* 68 (2015): 53–64.

26. C. Oren, L. Peled-Avron, and S. G. Shamay-Tsoory, "A Scent of Romance: Human Putative Pheromone Affects Men's Sexual Cognition," *Social Cognitive and Affective Neuroscience* 14 (2019): 719–26.

27. W. Zhou et al., "Chemosensory Communication of Gender Through Two Human Steroids in a Sexually Dimorphic Manner," *Current Biology* 24 (2014): 1091–95.

28. R. M. Hare et al., "Putative Sex-Specific Human Pheromones Do Not Affect Gender Perception, Attractiveness Ratings or Unfaithfulness Judgements of Opposite Sex Faces," *Royal Society Open Science* 4 (2017): 160831, https://doi.org/10.1098/rsos.160831.

29. H. Leder, A. Mitrovic, and J. Goller, "How Beauty Determines Gaze! Facial Attractiveness and Gaze Duration in Images of Real World Scenes," *Iperception* 7 (2016): 2041669516664355, https://doi.org /10.1177/2041669516664355; R. Lippa, T. M. Patterson, and W. D. Marelich, "Looking at and Longing for Male and Female 'Swimsuit Models': Men Are Much More Category-Specific Than Women," *Social Psychological and Personality Science* 1 (2010): 238–45.

30. G. Rieger and R. C. Savin-Williams, "The Eyes Have It: Sex and Sexual Orientation Differences in Pupil Dilation Patterns," *PLoS One* 7 (2012): e40256, https://doi.org/10.1371/journal.pone.0040256.

31. R. A. Hoss and J. H. Langlois, "Infants Prefer Attractive Faces," in *The Development of Face Processing in Infancy and Early Childhood: Current Perspectives*, ed. O. Pascalis and A. Slater (New York: Nova Science, 2003).

3. AROUSAL

1. D. Blumer, "Hypersexual Episodes in Temporal Lobe Epilepsy," *American Journal of Psychiatry* 126 (1970): 83–90.

2. L. Chaton et al., "Localization of an Epileptic Orgasmic Feeling to the Right Amygdala, Using Intracranial Electrodes," *Cortex* 109 (2018): 347–51.

3. C. C. Carbon et al., "First Gender, Then Attractiveness: Indications of Gender-Specific Attractiveness Processing via ERP Onsets," *Neuroscience Letters* 686 (2018): 186–92.

4. E. Mitricheva et al., "Neural Substrates of Sexual Arousal Are Not Sex Dependent," *PNAS* 116 (2019): 15671–76.

5. J. W. Seok, M. S. Park, and J. H. Sohn, "Neural Pathways in Processing of Sexual Arousal: A Dynamic Causal Modeling Study," *International Journal of Impotence Research* 28 (2016): 184–88.

6. M. Parada et al., "How Hot Are They? Neural Correlates of Genital Arousal: An Infrared Thermographic and Functional Magnetic Resonance Imaging Study of Sexual Arousal in Men and Women," *Journal of Sexual Medicine* 15 (2018): 217–29.

7. M. P. Richardson, B. A. Strange, and R. J. Dolan, "Encoding of Emotional Memories Depends on Amygdala and Hippocampus and Their Interactions," *Nature Neuroscience* 7 (2004): 278–85; F. Dolcos et al., "Emerging Directions in Emotional Episodic Memory," *Frontiers in Psychology* 8 (2017): 1867.

8. M. Poletti, C. Lucetti, and U. Bonuccelli, "Out-of-Control Sexual Behavior in an Orbitofrontal Cortex-Damaged Elderly Patient," *Journal of Neuropsychiatry and Clinical Neurosciences* 22 (2010): E7.

9. A. J. Pegna et al., "Discriminating Emotional Faces Without Primary Visual Cortices Involves the Right Amygdala," *Nature Neuroscience* 8 (2005): 24–25.

10. M. Wernicke et al., "Neural Correlates of Subliminally Presented Visual Sexual Stimuli," *Consciousness and Cognition* 49 (2017): 35–52.

11. O. Gillath and T. Collins, "Unconscious Desire: The Affective and Motivational Aspects of Subliminal Sexual Priming," *Archives of Sexual Behavior* 45 (2016): 5–20.

12. Y. Jiang et al., "A Gender- and Sexual Orientation-Dependent Spatial Attentional Effect of Invisible Images," *PNAS* 103 (2006): 17048–52.

13. I. Savic et al., "Smelling of Odorous Sex Hormone-Like Compounds Causes Sex-Differentiated Hypothalamic Activations in Humans," *Neuron* 31 (2001): 661–68.

14. S. M. Burke et al., "Heterosexual Men and Women Both Show a Hypothalamic Response to the Chemo-Signal Androstadienone," *PLoS One* 7: e40993, https://doi.org/10.1371/journal.pone.0040993.

15. T. Yamaguchi et al., "Posterior Amygdala Regulates Sexual and Aggressive Behaviors in Male Mice," *Nature Neuroscience* 23 (2020): 1111–24.

16. D. W. Bayless et al., "Limbic Neurons Shape Sex Recognition and Social Behavior in Sexually Naive Males," *Cell* 176 (2019): 1190–1205.

17. Y. Y. Fang et al., "A Hypothalamic-Midbrain Pathway Essential for Driving Maternal Behaviors," *Neuron* 98 (2018): 192–207.

18. H. K. Huynh et al., "Pontine Control of Ejaculation and Female Orgasm," *Journal of Sexual Medicine* 10 (2013): 3038–48.

19. B. F. Blok, A. T. Willemsen, and G. Holstege, "A PET Study on Brain Control of Micturition in Humans," *Brain* 120, pt. 1 (1997): 111–21.

20. A. Nowogrodzki, "The World's Strongest MRI Machines Are Pushing Human Imaging to New Limits," *Nature* 563 (2018): 24–26.

4. ORIENTATION

1. H. Wofford, "Finding Love Again, This Time with a Man," *New York Times*, April 23, 2016.

2. K. D. Suschinsky and M. L. Lalumiere, "Prepared for Anything?: An Investigation of Female Genital Arousal in Response to Rape Cues," *Psychological Science* 22 (2011): 159–65.

3. Centers for Disease Control, National Center for Health Statistics, "Key Statistics from the National Survey of Family Growth—S Listing," https://www.cdc.gov/nchs/nsfg/key_statistics/s.htm#sexualattraction.

4. D. Compton and T. Bridges, "2018 GSS Update on the U.S. LGB Population," Inequality by (Interior) Design (blog), April 12, 2019, https://inequalitybyinteriordesign.wordpress.com/2019/04/12/2018-gss-update-on-the-u-s-lgb-population/.

5. S. E. Mock and R. P. Eibach, "Stability and Change in Sexual Orientation Identity Over a 10-Year Period in Adulthood," *Archives of Sexual Behavior* 41 (2012): 641–48.

6. L. M. Diamond, *Sexual Fluidity: Understanding Women's Love and Desire* (Cambridge, MA: Harvard University Press, 2008).

7. D. C. Gruia et al., "Stability and Change in Sexual Orientation and Genital Arousal Over Time," *Journal of Sex Research* (April 12, 2022), https://doi.org/10.1080/00224499.2022.2060927.

8. A. M. Rosenthal et al., "The Male Bisexuality Debate Revisited: Some Bisexual Men Have Bisexual Arousal Patterns," *Archives of Sexual Behavior* 41 (2012): 135–47.

9. K. D. Suschinsky et al., "Use of the Bogus Pipeline Increases Sexual Concordance in Women But Not Men," *Archives of Sexual Behavior* 49 (2020): 1517–32.

10. M. L. Chivers, "The Specificity of Women's Sexual Response and Its Relationship with Sexual Orientations: A Review and Ten Hypotheses," *Archives of Sexual Behavior* 46 (2017): 1161–79.

11. R. A. Lippa, "Category Specificity of Self-Reported Sexual Attraction and Viewing Times to Male and Female Models in a Large U.S. Sample: Sex, Sexual Orientation, and Demographic Effects," *Archives of Sexual Behavior* 46 (2017): 167–78.

12. A. Ganna et al., "Large-Scale GWAS Reveals Insights Into the Genetic Architecture of Same-Sex Sexual Behavior," *Science* 365 (2019): 882.

13. Jones, J. M., "LGBT Identification Rises to 5.6 Percent in Latest U.S. Estimate," Gallup, February 24, 2021, https://news.gallup.com/poll/329708/lgbt-identification-rises-latest-estimate.aspx.

14. M. Apostolou, "The Direct Reproductive Cost of Same-Sex Attraction: Evidence from Two Nationally Representative U.S. Samples," *Archives of Sexual Behavior* 51 (2022): 1857–64.

15. D. P. VanderLaan, L. J. Petterson, and P. L. Vasey, "Elevated Kin-Directed Altruism Emerges in Childhood and Is Linked to Feminine Gender Expression in Samoan *Fa'afafine*: A Retrospective Study," *Archives of Sexual Behavior* 46 (2017): 95–108.

16. A. Camperio-Ciani, F. Corna, and C. Capiluppi, "Evidence for Maternally Inherited Factors Favouring Male Homosexuality and Promoting Female Fecundity," *Proceedings of the Royal Society of London. Series B: Biological Sciences* 271 (2004): 2217–21.

17. S. LeVay, *Gay, Straight, and the Reason Why: The Science of Sexual Orientation*, 2nd ed. (Oxford: Oxford University Press, 2017).

18. R. A. Gorski et al., "Evidence for a Morphological Sex Difference Within the Medial Preoptic Area of the Rat Brain," *Brain Research* 148 (1978): 333–46.

19. R. M. Jordan-Young, *Brain Storm: The Flaws in the Science of Sex Differences* (Cambridge, MA: Harvard University Press, 2011).

20. L. Eliot et al., "Dump the 'Dimorphism': Comprehensive Synthesis of Human Brain Studies Reveals Few Male-Female Differences Beyond Size," *Neuroscience and Biobehavioral Reviews* 125 (2021): 667–97.

21. S. LeVay, "A Difference in Hypothalamic Structure Between Heterosexual and Homosexual Men," *Science* 253 (1991): 1034–37.

22. W. Byne et al., "The Interstitial Nuclei of the Human Anterior Hypothalamus: An Investigation of Variation with Sex, Sexual Orientation, and HIV Status," *Hormones and Behavior* 40 (2001): 86–92.

23. C. E. Roselli et al., "The Volume of a Sexually Dimorphic Nucleus in the Ovine Medial Preoptic Area/Anterior Hypothalamus Varies with Sexual Partner Preference," *Endocrinology* 145 (2004): 478–83.

24. M. S. Allen and D. A. Robson, "Personality and Sexual Orientation: New Data and Meta-analysis," *Journal of Sex Research* 57 (2020): 953–65.

25. F. P. Kruijver et al., "Male-to-Female Transsexuals Have Female Neuron Numbers in a Limbic Nucleus," *Journal of Clinical Endocrinology and Metabolism* 85 (2000): 2034–41.

26. C. Abe et al., "Cross-Sex Shifts in Two Brain Imaging Phenotypes and Their Relation to Polygenic Scores for Same-Sex Sexual Behavior: A Study of 18,645 Individuals from the UK Biobank," *Human Brain Mapping* 42 (2021): 2292–304.

27. T. Paul et al., "Brain Response to Visual Sexual Stimuli in Heterosexual and Homosexual Males," *Human Brain Mapping* 29 (2008): 726–35.

28. I. Savic, H. Berglund, and P. Lindstrom, "Brain Response to Putative Pheromones in Homosexual Men," *PNAS* 102 (2005): 7356–61; H. Berglund, P. Lindstrom, and I. Savic, "Brain Response to Putative Pheromones in Lesbian Women," *PNAS* 103 (2006): 8269–74.

29. W. M. Brown et al., "Masculinized Finger Length Patterns in Human Males and Females with Congenital Adrenal Hyperplasia," *Hormones and Behavior* 42 (2002): 380–86.

30. T. Grimbos et al., "Sexual Orientation and the Second to Fourth Finger Length Ratio: A Meta-Analysis in Men and Women," *Behavioral*

Neuroscience 124 (2010): 278–87; S. M. Breedlove, "Replicable Data for Digit Ratio Differences," *Science* 365 (2019): 230.

31. J. V. Valentova et al., "Shape Differences Between the Faces of Homosexual and Heterosexual Men," *Archives of Sexual Behavior* 43 (2014): 353–61.

32. Y. Wang and M. Kosinski, "Deep Neural Networks Are More Accurate Than Humans at Detecting Sexual Orientation from Facial Images," *Journal of Personality and Social Psychology* 114 (2018): 246–57.

33. M. N. Skorska et al., "Facial Structure Predicts Sexual Orientation in Both Men and Women," *Archives of Sexual Behavior* 44 (2015): 1377–94.

34. A. F. Bogaert and S. Hershberger, "The Relation Between Sexual Orientation and Penile Size," *Archives of Sexual Behavior* 28 (1999): 213–21.

35. R. Blanchard, "Fraternal Birth Order, Family Size, and Male Homosexuality: Meta-Analysis of Studies Spanning 25 Years," *Archives of Sexual Behavior* 47 (2018): 1–15.

36. J. K. Vilsmeier et al., "The Fraternal Birth-Order Effect as Statistical Artefact: Convergent Evidence from Probability Calculus, Simulated Data, and Multiverse Meta-Analysis," PsyArXiv, May 28, 2021, doi:10.31234/osf.io/e4j6a.

37. A. F. Bogaert, "Biological Versus Nonbiological Older Brothers and Men's Sexual Orientation," *PNAS* 103 (2006): 10771–74.

38. A. F. Bogaert et al., "Male Homosexuality and Maternal Immune Responsivity to the Y-Linked Protein NLGN4Y," *PNAS* 115 (2018): 302–6.

39. R. Gondim, F. Teles, and U. Barroso, Jr., "Sexual Orientation of 46, XX Patients with Congenital Adrenal Hyperplasia: A Descriptive Review," *Journal of Pediatric Urology* 14 (2018): 486–93.

40. T. C. Ngun et al., "A Novel Predictive Model of Sexual Orientation Using Epigenetic Markers" (presented at the in American Society of Human Genetics Annual Meeting, Baltimore, MD, 2015).

41. S. M. Breedlove, "Prenatal Influences on Human Sexual Orientation: Expectations Versus Data," *Archives of Sexual Behavior* 46 (2017): 1583–92.

42. W. M. Brown et al., "Differences in Finger Length Ratios Between Self-Identified 'Butch' and 'Femme' Lesbians," *Archives of Sexual Behavior* 31 (2002): 123–27.

43. A. Swift-Gallant et al., "Differences in Digit Ratios Between Gay Men Who Prefer Receptive Versus Insertive Sex Roles Indicate a Role for Prenatal Androgen," *Scientific Reports* 11 (2021): 8102.

5. HAVING SEX

1. M. A. Killingsworth and D. T. Gilbert, "A Wandering Mind Is an Unhappy Mind," *Science* 330 (2010): 932.
2. A. L. Nelson and C. Purdon, "Non-Erotic Thoughts, Attentional Focus, and Sexual Problems in a Community Sample," *Archives of Sexual Behavior* 40 (2011): 395–406.
3. S. H. Oakley et al., "Clitoral Size and Location in Relation to Sexual Function Using Pelvic MRI," *Journal of Sexual Medicine* 11 (2014): 1013–22.
4. S. Aydin et al., "The Role of Clitoral Topography in Sexual Arousal and Orgasm: Transperineal Ultrasound Study," *International Urogynecology Journal* 33 (2022): 1495–1502.
5. K. Wallen. and E. A. Lloyd, "Female Sexual Arousal: Genital Anatomy and Orgasm in Intercourse," *Hormones and Behavior* 59 (2011): 780–92.
6. E. W. Eichel, J. D. Eichel, and S. Kule, "The Technique of Coital Alignment and Its Relation to Female Orgasmic Response and Simultaneous Orgasm," *Journal of Sex and Marital Therapy* 14 (1988): 129–41.
7. R. J. Levin, "The Clitoral Activation Paradox—Claimed Outcomes from Different Methods of Its Stimulation," *Clinical Anatomy* 31 (2018): 650–60.
8. E. A. Lloyd, *The Case of the Female Orgasm: Bias in the Science of Evolution* (Cambridge, MA: Harvard University Press, 2006).
9. A. K. Ladas, B. Whipple, and J. D. Perry, *The G Spot and Other Recent Discoveries About Human Sexuality* (New York: Holt, 1982).
10. B. Whipple, C. A. Gerdes, and B. R. Komisaruk, "Sexual Response to Self-Stimulation in Women with Complete Spinal Cord Injury," *Journal of Sex Research* 33 (1996): 231–40.
11. Anonymous, "Cervical Orgasms Are Incredible," Reddit, accessed August 19, 2022, https://www.reddit.com/r/sex/comments/95fjmv/cervical_orgasms_are_incredible/.
12. A. Barrell, "What Is Female Ejaculation?", *Medical News Today*, January 20, 2020, https://www.medicalnewstoday.com/articles/323953.

13. G. Schubach, "Urethral Expulsions During Sensual Arousal and Bladder Catheterization in Seven Human Females," *Electronic Journal of Human Sexuality* (defunct) 4 (August 25, 2001), http://www.ejhs.org /volume4/Schubach/Intro.html..

14. G. De Cuypere et al., "Sexual and Physical Health After Sex Reassignment Surgery," *Archives of Sexual Behavior* 34 (2005): 679–90.

15. E. B. Vance and N. N. Wagner, "Written Descriptions of Orgasms: A Study of Sex Differences," *Archives of Sexual Behavior* 5 (1976): 87–98.

16. M. Gerard et al., "Female Multiple Orgasm: An Exploratory Internet-Based Survey," *Journal of Sex Research* 58 (2021): 206–21.

17. B. Whipple, B. Myers, and B. R. Komisaruk, "Male Multiple Ejaculatory Orgasms: A Case Study," *Journal of Sex Education and Therapy* 23 (1998): 157–62.

18. P. Haake et al., "Absence of Orgasm-Induced Prolactin Secretion in a Healthy Multi-Orgasmic Male Subject," *International Journal of Impotence Research* 14 (2002): 133–35.

19. T. H. Kruger et al., "Effects of Acute Prolactin Manipulation on Sexual Drive and Function in Males," *Journal of Endocrinology* 179 (2003): 357–65.

20. A. B. Hollander et al., "Cabergoline in the Treatment of Male Orgasmic Disorder: A Retrospective Pilot Analysis," *Journal of Sexual Medicine* 4 (2016): e28–e33, https://doi.org/10.1016/j.esxm.2015.09.001.

21. Center for Sexual Health Promotion, Indiana University School of Public Health, "About the National Survey of Sexual Health and Behavior," https://nationalsexstudy.indiana.edu/.

22. R. J. Levin, "Prostate-Induced Orgasms: A Concise Review Illustrated with a Highly Relevant Case Study," *Clinical Anatomy* 31 (2018): 81–85.

23. D. Bohlen et al., "Five Meters of H(2)O: The Pressure at the Urinary Bladder Neck During Human Ejaculation," *Prostate* 44 (2000): 339–41.

24. G. G. Gallup, R. L. Burch, and T. J. Mitchell, "Semen Displacement as a Sperm Competition Strategy: Multiple Mating, Self-Semen Displacement, and Timing of In-Pair Copulations," *Human Nature* 17 (2006): 253–64.

25. M. Spehr et al., "Identification of a Testicular Odorant Receptor Mediating Human Sperm Chemotaxis," *Science* 299 (2003): 2054–58.

26. L. Larsson and M. Laska, "Ultra-High Olfactory Sensitivity for the Human Sperm-Attractant Aromatic Aldehyde Bourgeonal in CD-1 Mice," *Neuroscience Research* 71 (2011): 355–60.

27. G. Ottaviano et al., "Human Olfactory Sensitivity for Bourgeonal and Male Infertility: A Preliminary Investigation," *European Archives of Oto-Rhino-Laryngology* 270 (2013): 3079–86.

28. H. S. Kaplan, *Disorders of Sexual Desire and Other New Concepts and Techniques in Sex Therapy* (New York: Simon and Schuster, 1979).

29. M. A. Yule, L. A. Brotto, and B. B. Gorzalka, "Human Asexuality: What Do We Know About a Lack of Sexual Attraction?", *Current Sexual Health Reports* 9 (2017): 50–56.

30. R. Spencer, Libidos, Vibrators and Men, Oh My! This Is What Your Ageing Sex Drive Looks Like," *Guardian*, March 25, 2014.

31. C. M. Meston and D. M. Buss, *Why Women Have Sex: The Psychology of Sex in Women's Own Voices* (New York: Times Books, 2009).

32. L. Tiefer, "The New 'Female Sexual Dysfunction': A Story of Medicalization, Disease-Mongering, and Resistance" (video), Tiefer Talk IU, Indiana University, https://iu.mediaspace.kaltura.com/media/Tiefer+Talk +IU/1_wiqs4bjz.

33. N. E. Avis et al., "Change in Sexual Functioning Over the Menopausal Transition: Results from the Study of Women's Health Across the Nation," *Menopause* 24 (2017): 379–90.

34. M. Cappelletti and K. Wallen, "Increasing Women's Sexual Desire: The Comparative Effectiveness of Estrogens and Androgens," *Hormones and Behavior* 78 (2016): 178–93.

35. H. G. Nurnberg et al., "Sildenafil Treatment of Women with Antidepressant-Associated Sexual Dysfunction: A Randomized Controlled Trial," *JAMA* 300 (2008): 395–404.

36. M. E. Hadley, "Discovery That a Melanocortin Regulates Sexual Functions in Male and Female Humans," *Peptides* 26 (2005): 1687–89.

37. S. A. Kingsberg et al., "Bremelanotide for the Treatment of Hypoactive Sexual Desire Disorder: Two Randomized Phase 3 Trials," *Obstetrics and Gynecology* 134 (2019): 899–908.

38. L. Jaspers et al., "Efficacy and Safety of Flibanserin for the Treatment of Hypoactive Sexual Desire Disorder in Women: A Systematic Review and Meta-Analysis," *JAMA Internal Medicine* 176 (2016): 453–62.

39. L. Habbema et al., "Risks of Unregulated Use of Alpha-Melanocyte-Stimulating Hormone Analogues: A Review," *International Journal of Dermatology* 56 (2017): 975–80.

40. G. I. Spielmans, "Re-Analyzing Phase III Bremelanotide Trials for 'Hypoactive Sexual Desire Disorder' in Women," *Journal of Sex Research* 58 (2021): 1085–1105.

41. R. Basson, "Human Sexual Response," in *Handbook of Clinical Neurology, vol. 130 (3rd series): Neurology of Sexual and Bladder Disorders*, ed. D. B. Vodusek and F. Boller (Amsterdam: Elsevier, 2015), 11–18.

42. R. Basson, M. Driscoll, and S. Correia, "Flibanserin for Low Sexual Desire in Women: A Molecule from Bench to Bed?", *EBioMedicine* 2 (2015): 772–73.

6. RELATIONSHIPS

1. R. Pande, *Learning to Love: Arranged Marriages and the British Indian Diaspora* (New Brunswick, NJ: Rutgers University Press, 2021).

2. M. J. Rosenfeld, R. J. Thomas, and S. Hausen, "Disintermediating Your Friends: How Online Dating in the United States Displaces Other Ways of Meeting," *PNAS* 116 (2019): 17753–58.

3. A. Paul, "Is Online Better Than Offline for Meeting Partners? Depends: Are You Looking to Marry or to Date?", *Cyberpsychology, Behavior, and Social Networking* 17 (2014): 664–67.

4. R. Kurzban and J. Weeden, "HurryDate: Mate Preferences in Action," *Evolution and Human Behavior* 26 (2005): 227–44.

5. E. P. Eastwick and E. J. Finkel, "Sex Differences in Mate Preferences Revisited: Do People Know What They Initially Desire in a Romantic Partner?", *Journal of Personality and Social Psychology* 94 (2008): 245–64.

6. D. A. Stinson, J. J. Cameron, and L. B. Hoplock, "The Friends-to-Lovers Pathway to Romance: Prevalent, Preferred, and Overlooked by Science," *Social Psychological and Personality Science* 13 (2022): 562–71.

7. C. McLaren, "What Dating Is Like as a Gay Deaf Man," Logo, August 14, 2017, https://www.logotv.com/news/up7hn5/gay-deaf-dating.

8. P. W. Eastwick and L. L. Hunt, "Relational Mate Value: Consensus and Uniqueness in Romantic Evaluations," *Journal of Personality and Social Psychology* 106 (2014): 728–51.

9. E. Hatfield, G. W. Walster, and E. Berscheid, *Equity Theory and Research* (Boston: Allyn & Bacon, 1978).

10. F. Kocsor et al., "Preference for Facial Self-Resemblance and Attractiveness in Human Mate Choice," *Archives of Sexual Behavior* 40 (2011): 1263–70.

11. I. Croy et al., "Marriage Does Not Relate to Major Histocompatibility Complex: A Genetic Analysis Based on 3691 Couples," *Proceedings of the Royal Society of London. Series B: Biological Sciences* 287 (2020): 20201800.

12. J. Kromer et al., "Influence of HLA on Human Partnership and Sexual Satisfaction," *Scientific Reports* 6 (2016): 32550.

13. E. M. Giolla and P. J. Kajonius, "Sex Differences in Personality Are Larger in Gender Equal Countries: Replicating and Extending a Surprising Finding," *International Journal of Psychology* 54 (2019): 705–11.

14. C. R. Schwartz and N. L. Graf, "Assortative Matching Among Same-Sex and Different-Sex Couples in the United States, 1990–2000," *Demographic Research* 21 (2009): 843–78.

15. M. Develin, "The Age of Love," https://m.facebook.com/nt/screen/?params=%7B%22note_id%22%3A101589288005158415%7D&path=%2Fnotes%2Fnote%2F&_rdr.

16. D. A. Moskowitz and T. A. Hart, "The Influence of Physical Body Traits and Masculinity on Anal Sex Roles in Gay and Bisexual Men," *Archives of Sexual Behavior* 40 (2011): 835–41.

17. Aristotle, *Nicomachean Ethics*, book 8, trans. W. D. Ross (Ontario, Canada: Batoche, 1999).

18. S. Leikas et al., "Relationship Satisfaction and Similarity of Personality Traits, Personal Values, and Attitudes," *Personality and Individual Differences* 123 (2018): 191–98.

19. M. Shiota and R. W. Levenson, "Birds of a Feather Don't Always Fly Farthest: Similarity in Big Five Personality Predicts More Negative Marital Satisfaction Trajectories in Long-Term Marriages," *Psychology and Aging* 22 (2007): 666–75.

20. P. Gonalon-Pons and C. R. Schwartz, "Trends in Economic Homogamy: Changes in Assortative Mating or the Division of Labor in Marriage?", *Demography* 54 (2017): 985–1005.

21. R. L. Hopcroft, "High Income Men Have High Value as Long-Term Mates in the U.S.: Personal Income and the Probability of Marriage, Divorce, and Childbearing in the U.S.," *Evolution and Human Behavior* 42 (2021): 409–17.

22. Y. Qian, "Gender Asymmetry in Educational and Income Assortative Marriage," *Journal of Marriage and Family* 79 (2017): 318–36.

23. J. M. Gottman and R. W. Levenson, "The Timing of Divorce: Predicting When a Couple Will Divorce Over a 14-Year Period," *Journal of Marriage and the Family* 62 (2000): 737–45.

24. Gottman Institute, "Love Lab: The Original Couples Laboratory, Reimagined for the 21st Century," https://www.gottman.com/love-lab/.

25. J. M. Gottman and N. Silver, *The Seven Principles for Making Marriage Work*, 2nd rev. ed. (New York: Harmony, 2015).

26. J. A. Lavner, B. R. Karney, and T. N. Bradbury, "Does Couples' Communication Predict Marital Satisfaction, or Does Marital Satisfaction Predict Communication?", *Journal of Marriage and Family* 78 (2016): 680–94.

27. M. D. Johnson et al., "Within-Couple Associations Between Communication and Relationship Satisfaction Over Time," *Personality and Social Psychology Bulletin* 48 (2022): 534–49.

28. J. Gottman, "The Empirical Basis for Gottman Method Couples Therapy," Gottman Institute, https://www.gottman.com/blog/the-empirical -basis-for-gottman-method-couples-therapy/.

29. R. G. Wood et al., "The Building Strong Families Project," May 2010, https://web.archive.org/web/20130720101532/http://www.mathematica -mpr.com/publications/pdfs/family_support/BSF_impact_finalrpt.pdf.

30. D. M. Buss et al., "Sex Differences in Jealousy: Evolution, Physiology, and Psychology," *Psychological Science* 3 (1992): 251–55.

31. D. M. Buss, "Sexual and Emotional Infidelity: Evolved Gender Differences in Jealousy Prove Robust and Replicable," *Perspectives in Psychological Science* 13 (2018): 155–60.

32. P. Dijkstra and B. P. Buunk, "Jealousy as a Function of Rival Characteristics: An Evolutionary Perspective," *Personality and Social Psychology Bulletin* 24 (1998): 1158–66.

33. T. V. Pollet and T. K. Saxton, "Jealousy as a Function of Rival Characteristics: Two Large Replication Studies and Meta-Analyses Support Gender Differences in Reactions to Rival Attractiveness But Not Dominance," *Personality and Social Psychology Bulletin* 46 (2020): 1428–43.

34. J. V. Valentova, A. C. de Moraes, and M. A. C. Varella, "Gender, Sexual Orientation and Type of Relationship Influence Individual Differences in Jealousy: A Large Brazilian Sample," *Personality and Individual Differences* 157 (2020), https://doi.org/10.1016/j.paid.2019.109805.

35. J. E. Edlund and B. J. Sagarin, "Sex Differences in Jealousy: A 25-Year Retrospective," *Advances in Experimental Social Psychology* 55 (2017): 259–302.

7. PARAPHILIAS

1. J. P. Fedoroff, *The Paraphilias: Changing Suits in the Evolution of Sexual Interest Paradigms* (Oxford: Oxford University Press, 2020).
2. A. Aggrawal, *Forensic and Medico-Legal Aspects of Sexual Crimes and Unusual Sexual Practices* (Boca Raton, FL: CRC Press, 2008).
3. K. Gates, *Deviant Desires: A Tour of the Erotic Edge* (New York: power-House, 2017).
4. S. Toska, "I Worked as a Dominatrix for Over 5 Years. Here's What It's Really Like," *Huffington Post*, March 6, 2019, https://www.huffpost.com/entry/working-as-a-dominatrix_n_5c66ea02e4b033a799423973.
5. A. Madill and Y. Zhao, "Are Female Paraphilias Hiding in Plain Sight? Risqué Male-Male Erotica for Women in Sinophone and Anglophone Regions," *Archives of Sexual Behavior* 51 (2022): 897–910.
6. C. C. Joyal, A. Cossette, and V. Lapierre, "What Exactly Is an Unusual Sexual Fantasy?", *Journal of Sexual Medicine* 12 (2015): 328–40.
7. M. S. Weinberg, C. J. Williams, and C. Calhan, "If the Shoe Fits . . . : Exploring Male Homosexual Foot Fetishism," *Journal of Sex Research* 32 (1995): 17-27.
8. J. S. Hyde and J. D. DeLamater, *Understanding Human Sexuality*, 7th ed. (New York: McGraw Hill, 2000).
9. S. Rachman, "Sexual Fetishism: An Experimental Analogue," *Psychological Record* 16 (1966): 293–96.
10. J. G. Pfaus et al., "Conditioning of Sexual Interests and Paraphilias in Humans Is Difficult to See, Virtually Impossible to Test, and Probably Exactly How It Happens: A Comment on Hsu and Bailey (2020)," *Archives of Sexual Behavior* 49 (2020): 1403–7.
11. G. G. Abel and C. A. Osborn, "The Paraphilias," in *New Oxford Textbook of Psychiatry*, ed. M. G. Gelder, J. J. López-Ibor, and N. Andreasen (Oxford: Oxford University Press, 2000), 817–913.
12. K. Alanko et al., "Evidence for Heritability of Adult Men's Sexual Interest in Youth Under Age 16 from a Population-Based Extended Twin Design," *Journal of Sexual Medicine* 10 (2013): 1090–99.

13. K. J. Hsu and J. M. Bailey, "The 'Furry' Phenomenon: Characterizing Sexual Orientation, Sexual Motivation, and Erotic Target Identity Inversions in Male Furries," *Archives of Sexual Behavior* 48 (2019): 1349–69.

14. D. W. Soh and J. M. Cantor, "A Peek Inside a Furry Convention," *Archives of Sexual Behavior* 44 (2015): 1–2.

15. K. J. Zucker and R. Blanchard, "Transvestic Fetishism: Psychopathology and Theory," in *Sexual Deviance: Theory, Assessment, and Treatment*, 2nd ed., ed. D. R. Laws and W. T. O'Donohue (New York: Guilford, 1997), 272–84.

16. R. Blanchard, "Clinical Observations and Systematic Studies of Autogynephilia," *Journal of Sex and Marital Therapy* 17 (1991): 235–51.

17. J. M. Serano, "The Case Against Autogynephilia," *International Journal of Transgenderism* 12 (2010): 176–87.

18. A. A. Lawrence, *Men Trapped in Men's Bodies: Narratives of Autogynephilic Transsexualism* (New York: Springer, 2013).

19. A. A. Lawrence, "Erotic Target Location Errors: An Underappreciated Paraphilic Dimension," *Journal of Sex Research* 46 (2009): 194–215.

20. M. Bailey, "Michael Jackson: Erotic Identity Disorder?", Science 2.0, July 1, 2009, https://www.science20.com/j_michael_bailey/michael_jackson_erotic_identity_disorder-55152.

21. A. C. Eberhart, *There Goes Sunday School* (Jupiter, FL: Seven Sisters, 2020)

22. G. Schmidt, ed., *Kinder der Sexuellen Revolution* (Gießen, Germany: Psychosozial-Verlag, 2000).

23. H. Morton and B. B. Gorzalka, "Role of Partner Novelty in Sexual Functioning: A Review," *Journal of Sex and Marital Therapy* 41 (2015): 593–609.

24. C. Carlström, "Spiritual Experiences and Altered States of Consciousness—Parallels Between BDSM and Christianity," *Sexualities* 24 (2021): 749–66.

8. PEDOPHILIA

1. R. Lievesley and R. Lapworth, "'We Do Exist': The Experiences of Women Living with a Sexual Interest in Minors," *Archives of Sexual Behavior* 51 (2022): 879–96.

2. C. Leach, A. Stewart, and S. Smallbone, "Testing the Sexually Abused-Sexual Abuser Hypothesis: A Prospective Longitudinal Birth Cohort Study," *Child Abuse and Neglect* 51 (2016): 144–53.

3. N. Papalia et al., "Child Sexual Abuse and Criminal Offending: Gender-Specific Effects and the Role of Abuse Characteristics and Other Adverse Outcomes," *Child Maltreatment* 23 (2018): 399–416.

4. S. Brown, *Against All Odds: My Life of Hardship, Fast Breaks, and Second Chances* (New York: Harper, 2011).

5. F. Dyshniku et al., "Minor Physical Anomalies as a Window Into the Prenatal Origins of Pedophilia," *Archives of Sexual Behavior* 44 (2015): 2151–59.

6. J. M. Cantor et al., "Diffusion Tensor Imaging of Pedophilia," *Archives of Sexual Behavior* 44 (2015): 2161–72.

7. C. Kargel et al., "Evidence for Superior Neurobiological and Behavioral Inhibitory Control Abilities in Non-Offending as Compared to Offending Pedophiles," *Human Brain Mapping* 38 (2017): 1092–1104.

8. S. Jahnke et al., "Neurodevelopmental Differences, Pedohebephilia, and Sexual Offending: Findings from Two Online Surveys," *Archives of Sexual Behavior* 51 (2022): 849–66.

9. A. F. Bogaert et al., "Pedophilia, Sexual Orientation, and Birth Order," *Journal of Abnormal Psychology* 106 (1997): 331–35.

10. A. Walker, "'I'm Not Like That, So Am I Gay?' The Use of Queer-Spectrum Identity Labels Among Minor-Attracted People," *Journal of Homosexuality* 67 (2020): 1736–59.

11. A. Hammer, "Johns Hopkins Child Sex Abuse Center Hires Trans Professor, 34, Who Was Forced to Resign from Virginia School for DEFENDING Pedophiles as 'Minor Attracted Persons,'" *Daily Mail*, May 13, 2022, https://www.dailymail.co.uk/news/article-10813233/Trans -professor-DEFENDED-pedophiles-minor-attracted-persons-hired -Johns-Hopkins-center.html.

12. M. C. Seto, "The Puzzle of Male Chronophilias," *Archives of Sexual Behavior* 46 (2017): 3–22.

13. F. M. Martijn, K. M. Babchishin, L. E. Pullman and M. C. Seto, "Sexual Attraction and Falling in Love in Persons with Pedohebephilia," *Archives of Sexual Behavior* 49 (2020): 1305–18.

14. "Welcome to Virtuous Pedophiles," Virtuous Pedophiles, accessed August 19, 2022, https://www.virped.org.

15. L. Cameron, "How the Psychiatrist Who Co-Wrote the Manual on Sex Talks About Sex," Vice, April 11, 2012, https://www.vice.com/en /article/ypp93m/heres-how-the-guy-who-wrote-the-manual-on-sex -talks-about-sex.

16. K. Müller et al., "Changes in Sexual Arousal as Measured by Penile Plethysmography in Men with Pedophilic Sexual Interest," *Journal of Sexual Medicine* 11 (2014): 1221–29.

17. A. Mokros and E. Habermeyer, "Regression to the Mean Mimicking Changes in Sexual Arousal to Child Stimuli in Pedophiles," *Archives of Sexual Behavior* 45 (2016): 1863–67.

18. I. V. McPhail and M. E. Olver, "Interventions for Pedohebephilic Arousal in Men Convicted for Sexual Offenses Against Children," *Criminal Justice and Behavior* 47 (2020): 1319–39.

19. J. P. Fedoroff et al., "Regression to the Mean or the Semmelweis Reflex?", *Archives of Sexual Behavior* 45 (2016): 1869–70.

20. T. Nickerson, "I'm a Pedophile, But Not a Monster," September 21, 2015, https://archive.md/ttVNy.

9. PORN

1. "The 2019 Year in Review," Pornhub, December 11, 2019, https://www.pornhub.com/insights/2019-year-in-review.

2. T. Kohut, W. A. Fisher, and L. Campbell, "Perceived Effects of Pornography on the Couple Relationship: Initial Findings of Open-Ended, Participant-Informed, 'Bottom-Up' Research," *Archives of Sexual Behavior* 46 (2017): 585–602.

3. K. Dawson et al., "Development of a Measure to Assess What Young Heterosexual Adults Say They Learn About Sex from Pornography," *Archives of Sexual Behavior* 51 (2022): 1257–69.

4. L. J. Seguin, C. Rodrigue, and J. Lavigne, "Consuming Ecstasy: Representations of Male and Female Orgasm in Mainstream Pornography," *Journal of Sex Research* 55 (2018): 348–56.

5. C. Innes, "Why Did We Ignore Porn's #MeToo?", *Cosmopolitan*, February 17, 2002, https://www.cosmopolitan.com/uk/reports/a30806689/james-deen-porn-me-too/.

6. E. J. Dickson, "Porn Star Asa Akira Takes Us Inside Her New Book, 'Insatiable,'" Daily Dot, May 20, 2014 (updated May 31, 2021), https://www.dailydot.com/irl/asa-akira-memoir-qanda/

7. "What Is a Day in the Life of a Porn Star Like?", Quora, accessed August 20, 2022, https://www.quora.com/What-is-a-day-in-the-life-of-a-porn-star-like-How-is-the-life-of-a-girl-different-from-a-guy,.

8. S. Harrison, "The Hard Life of the Male Porn Star," Your Tango, August 22, 2008, https://www.yourtango.com/20086446/the-hard-life -of-the-male-porn-star.

9. C. R. Grudzen et al., "Comparison of the Mental Health of Female Adult Film Performers and Other Young Women in California," *Psychiatric Services* 62 (2011): 639–45.

10. ABC News, "Paying a Lifelong Price for Porn Stint," May 27, 2004, https://abcnews.go.com/Primetime/story?id=132371&page=1.

11. A. Dworkin, *Letters from a War Zone* (New York: Dutton, 1989).

12. E. Mellor and S. Duff, "The Use of Pornography and the Relationship Between Pornography Exposure and Sexual Offending in Males: A Systematic Review," *Aggression and Violent Behavior* 46 (2019): 116–26.

13. W. L. Rostad et al., "The Association Between Exposure to Violent Pornography and Teen Dating Violence in Grade 10 High School Students," *Archives of Sexual Behavior* 48 (2019): 2137–47.

14. T. Kohut, I. Landripet, and A. Stulhofer, "Testing the Confluence Model of the Association Between Pornography Use and Male Sexual Aggression: A Longitudinal Assessment in Two Independent Adolescent Samples from Croatia," *Archives of Sexual Behavior* 50 (2021): 647–65.

15. S. G. Hatch et al., "Does Pornography Consumption Lead to Intimate Partner Violence Perpetration? Little Evidence for Temporal Precedence," *Canadian Journal of Human Sexuality* 29 (2020): 289–96.

16. S. Cunningham and M. Shah, "Decriminalizing Indoor Prostitution: Implications for Sexual Violence and Public Health" (NBER Working Paper No. 02138, National Bureau of Economic Research, Cambridge, MA, 2014), https://www.nber.org/system/files/working_papers /w20281/w20281.pdf.

17. R. Ciacci and M. M. Sviatschi, "The Effect of Adult Entertainment Establishments on Sex Crime: Evidence from New York City," *Economic Journal* 132 (2020): 147–98.

18. M. Diamond, E. Jozifkova, and P. Weiss, "Pornography and Sex Crimes in the Czech Republic," *Archives of Sexual Behavior* 40 (2011): 1037–43.

19. E. Shor and K. Seida, "'Harder and Harder'? Is Mainstream Pornography Becoming Increasingly Violent and Do Viewers Prefer Violent Content?", *Journal of Sex Research* 56 (2019): 16–28.

20. H. Stern, "Billie Eilish Opens Up About Surviving COVID and Hosting 'SNL,'" Howard Stern, December 13, 2021, https://www.howardstern

.com/news/2021/12/13/billie-eilish-performs-2-songs-live-in-studio
-and-opens-up-about-surviving-covid-and-hosting-snl/.

21. J. B. Grubbs, S. W. Kraus, and S. L. Perry, "Self-Reported Addiction to Pornography in a Nationally Representative Sample: The Roles of Use Habits, Religiousness, and Moral Incongruence," *Journal of Behavioral Addiction* 8 (2019): 88–93.

22. J. E. Grant et al., "Introduction to Behavioral Addictions," *American Journal of Drug and Alcohol Abuse* 36 (2010): 233–41.

23. P. Carnes, *Out of the Shadows: Understanding Sexual Addiction* (Center City, MN: Hazelden, 2001).

24. D. Ley, N. Prause, and P. Finn, "The Emperor Has No Clothes: A Review of the 'Pornography Addiction' Model," *Current Sexual Health Reports* 6 (2014): 94–105.

25. AutomaticEthos, "I've Reached My Breaking Point," Reddit, accessed August 20, 2022, https://www.reddit.com/r/NoFap/comments/pqkah6/ive_reached_my_breaking_point/.

26. B. Bőthe et al., "Are Sexual Functioning Problems Associated with Frequent Pornography Use and/or Problematic Pornography Use? Results from a Large Community Survey Including Males and Females," *Addictive Behaviors* 112 (2021): 106603, https://doi.org/10.1016/j.addbeh.2020.106603.

27. J. B. Grubbs et al., "Moral Incongruence and Compulsive Sexual Behavior: Results from Cross-Sectional Interactions and Parallel Growth Curve Analyses," *Journal of Abnormal Psychology* 129 (2020): 266–78.

28. S. Pappu, "Internet Porn Nearly Ruined His Life. Now He Wants to Help," *New York Times*, July 6, 2016.

29. VitalLampada, "From a Girl's Experience: Why Do Women Need to Be Here?", Reddit, accessed August 20, 2022, https://www.reddit.com/r/NoFap/comments/1yx6yz/from_a_girls_experience_why_do_women_need_to_be/.

30. D. Pirrone et al., "Pornography Use Profiles and the Emergence of Sexual Behaviors in Adolescence," *Archives of Sexual Behavior* 51 (2022): 1141–56.

31. D. T. Kenrick, S. E. Gutierres, and L. L. Goldberg, "Influence of Popular Erotica on Judgments of Strangers and Mates," *Journal of Experimental Social Psychology* 25 (1989): 159–67.

10. RAPE

1. T. Zerjal et al., "The Genetic Legacy of the Mongols," *American Journal of Human Genetics* 72 (2003): 717–21.

2. L. H. Wei et al., "Whole-Sequence Analysis Indicates That the Y Chromosome C2*-Star Cluster Traces Back to Ordinary Mongols, Rather Than Genghis Khan," *European Journal of Human Genetics* 26 (2018): 230–37.

3. J. Man, *Genghis Khan: Life, Death, and Resurrection* (New York: Thomas Dunne, 2004).

4. M. V. Derenko et al., "Distribution of the Male Lineages of Genghis Khan's Descendants in Northern Eurasian Populations," *Russian Journal of Genetics* 43 (2007): 334–37.

5. U. Onon, *The Secret History of the Mongols: The Life and Times of Chinggis Khan* (London: Routledge, 2011).

6. Y. Xue et al., "Recent Spread of a Y-Chromosomal Lineage in Northern China and Mongolia," *American Journal of Human Genetics* 77 (2005): 1112–16.

7. C. Yang et al., "The Association of 22 Y Chromosome Short Tandem Repeat Loci with Initiative-Aggressive Behavior," *Gene* 654 (2018): 10–13.

8. K. Bjorkqvist, "Gender Differences in Aggression," *Current Opinion in Psychology* 19 (2018): 39–42.

9. D. E. H. Russell, *Sexual Exploitation: Rape, Child Sexual Abuse, and Workplace Harassment* (Newbury Park, CA: SAGE, 1984).

10. B. A. McPhail, "Feminist Framework Plus: Knitting Feminist Theories of Rape Etiology Into a Comprehensive Model," *Trauma, Violence, & Abuse* 17 (2015): 314–29.

11. N. M. Malamuth and G. M. Hald, "The Confluence Mediational Model of Sexual Aggression," in *The Wiley Handbook on the Theories, Assessment, and Treatment of Sexual Offending*, ed D. P. Boer (New York: Wiley, 2017), 53–71.

12. S. N. Geniole et al., "Is Testosterone Linked to Human Aggression? A Meta-Analytic Examination of the Relationship Between Baseline, Dynamic, and Manipulated Testosterone on Human Aggression," *Hormones and Behavior* 123 (2020): 104644, https://doi.org/10.1016/j.yhbeh.2019.104644.

13. J. S. Wong and J. Gravel, "Do Sex Offenders Have Higher Levels of Testosterone? Results from a Meta-Analysis," *Sexual Abuse* 30 (2018): 147–68.

14. J. J. Turanovic and T. C. Pratt, "Exposure to Fetal Testosterone, Aggression, and Violent Behavior: A Meta-Analysis of the 2D:4D Digit Ratio," *Aggression and Violent Behavior* 33 (2017): 51–61.

15. S. da Cunha-Bang and G. M. Knudsen, "The Modulatory Role of Serotonin on Human Impulsive Aggression," *Biological Psychiatry* 90 (2021): 447–57.

16. H. G. Brunner et al., "Abnormal Behavior Associated with a Point Mutation in the Structural Gene for Monoamine Oxidase A," *Science* 262 (1993): 578–80.

17. O. Cases et al., "Aggressive Behavior and Altered Amounts of Brain Serotonin and Norepinephrine in Mice Lacking MAOA," *Science* 268 (1995): 1763–66.

18. N. J. Kolla and M. Bortolato, "The Role of Monoamine Oxidase A in the Neurobiology of Aggressive, Antisocial, and Violent Behavior: A Tale of Mice and Men," *Progress in Neurobiology* 194 (2020): 101875, https://doi.org/10.1016/j.pneurobio.2020.101875.

19. R. L. Trivers, "Parental Investment and Sexual Selection," in *Sexual Selection and the Descent of Man, 1871–1971*, ed. B. Campbell (Chicago: Aldine, 1972), 136–79.

20. N. Eldredge, *Why We Do It: Rethinking Sex and the Selfish Gene* (New York: Norton, 2004).

21. United Nations Population Division, "World Contraceptive Use 2022," https://www.un.org/development/desa/pd/data/world-contraceptive -use.

22. R. Thornhill and C. T. Palmer, *A Natural History of Rape: Biological Bases of Sexual Coercion* (Cambridge, MA: MIT Press, 2000).

23. P. R. Marty et al., "Endocrinological Correlates of Male Bimaturism in Wild Bornean Orangutans," *American Journal of Primatology* 77 (2015): 1170–78.

24. N. Malamuth, "An Evolutionary-Based Model Integrating Research on the Characteristics of Sexually Coercive Men," in *Advances in Psychological Science, Vol. 1: Personal, Social, and Developmental Aspects*, ed. J. G. Adair, D. Bélanger, and K. L. Dion (Hove, UK: Psychology Press, 1998), 151–84.

25. E. A. Smith, M. Borgerhoff Mulder, and K. Hill, "Controversies in the Evolutionary Social Sciences: A Guide for the Perplexed," *Trends in Ecology and Evolution* 16 (2001): 128–35.

26. Statista, "Number of Police Recorded Rape Offences in England and Wales from 2002/03 to 2020/21," accessed August 21, 2022, https://www.statista.com/statistics/283100/recorded-rape-offences-in-england-and-wales/.

27. C. J. Ryan, "Is It Really Ethical to Prescribe Antiandrogens to Sex Offenders to Decrease Their Risk of Recidivism?", in *Neurointerventions and the Law: Regulating Human Mental Capacity*, ed. N. A. Vincent, T. Nadelhoffer, and A. McCay (Oxford: Oxford University Press, 2020), 272–95.

28. M. Huppin, N. M. Malamuth, and D. Linz, "An Evolutionary Perspective on Sexual Assault and Implications for Interventions," in *Handbook of Sexual Assault and Sexual Assault Prevention*, ed. W. T. O'Donohue and P. A. Schewe (Cham, Switzerland: Springer, 2019), 17–44.

29. H. P. Hailes et al., "Long-Term Outcomes of Childhood Sexual Abuse: An Umbrella Review," *Lancet Psychiatry* 6 (2019): 830–39.

30. E. B. Foa and C. P. McLean, "The Efficacy of Exposure Therapy for Anxiety-Related Disorders and Its Underlying Mechanisms: The Case of OCD and PTSD," *Annual Review of Clinical Psychology* 12 (2016): 1–28.

31. P. Pace, *Lifespan Integration: Connecting Ego States Through Time* (Cle Elum, WA: Eirene Imprint, 2015).

32. G. Rajan et al., "A One-Session Treatment of PTSD After Single Sexual Assault Trauma. A Pilot Study of the WONSA MLI Project: A Randomized Controlled Trial," *Journal of Interpersonal Violence* 37 (2022): NP6582–NP6603, https://doi.org/10.1177/0886260520965973.

33. E. Smart and C. Stewart, *My Story* (New York: St. Martin's, 2013).

34. RAINN, "About Sexual Assault," accessed August 21, 2022, https://www.rainn.org/about-sexual-assault.

35. E. Teding van Berkhout and J. M. Malouff, "The Efficacy of Empathy Training: A Meta-Analysis of Randomized Controlled Trials," *Journal of Counsseling Psychology* 63 (2016): 32–41.

36. W. L. Marshall and L. E. Marshall, "Empathy and Sexual Offending: Theory, Research and Practice," in *Handbook of Sexual Assault and Sexual Assault Prevention*, ed. W. O'Donohue and P. Schewe (Cham, Switzerland: Springer, 2019), 229–39.

37. R. M. Leone, K. N. Oyler, and D. J. Parrott, "Empathy Is Not Enough: The Inhibiting Effects of Rape Myth Acceptance on the Relation Between Empathy and Bystander Intervention," *Journal of Interpersonal Violence* 36 (2021): 11532–52.

38. C. I. V. Bird, N. L. Modlin, and J. J. H. Rucker, "Psilocybin and MDMA for the Treatment of Trauma-Related Psychopathology," *International Review of Psychiatry* 33 (2021): 229–49.

11. LOVE

1. L. Getz, "Long-Term Vole Demographic Data Files," https://www.life.illinois.edu/getz/.

2. L. Getz, C. S. Carter, and L. Gavish, "The Mating System of the Prairie Vole, *Microtus ochrogaster*: Field and Laboratory Evidence for Pair-Bonding," *Behavioral Ecology and Sociobiology* 8 (1981): 189–94.

3. D. Lukas and T. H. Clutton-Brock, "The Evolution of Social Monogamy in Mammals," *Science* 341 (2013): 526–30.

4. J. B. Lichter et al., "Breeding Patterns of Female Prairie Voles (*Microtus ochrogaster*) Displaying Alternative Reproductive Tactics," *Journal of Mammalogy* 101 (2020): 990–99.

5. T. R. Insel, Z. X. Wang, and C. F. Ferris, "Patterns of Brain Vasopressin Receptor Distribution Associated with Social Organization in Microtine Rodents," *Journal of Neuroscience* 14 (1994): 5381–92.

6. M. M. Lim and L. J. Young, "Vasopressin-Dependent Neural Circuits Underlying Pair Bond Formation in the Monogamous Prairie Vole," *Neuroscience* 125 (2004): 35–45.

7. L. J. Young et al., "Increased Affiliative Response to Vasopressin in Mice Expressing the V1a Receptor from a Monogamous Vole," *Nature* 400 (1999): 766–68.

8. F. Duclot et al., "Transcriptomic Regulations Underlying Pair-Bond Formation and Maintenance in the Socially Monogamous Male and Female Prairie Vole," *Biological Psychiatry* 91 (2022): 141–51.

9. H. Walum and L. J. Young, "The Neural Mechanisms and Circuitry of the Pair Bond," *Nature Reviews of Neuroscience* 19 (2018): 643–54.

10. E. Fromm, *The Art of Loving* (New York: Harper & Row, 1956).

11. M. Shostak, *Nisa: The Life and Words of a !Kung Woman* (Cambridge, MA: Harvard University Press, 2000).

12. P. Sorokowski et al., "Universality of the Triangular Theory of Love: Adaptation and Psychometric Properties of the Triangular Love Scale in 25 Countries," *Journal of Sex Research* 58 (2021): 106–15.

13. K. L. Bales et al., "Titi Monkeys as a Novel Non-Human Primate Model for the Neurobiology of Pair Bonding," *Yale Journal of Biology and Medicine* 90 (2017): 373–87.

14. N. Scatliffe et al., "Oxytocin and Early Parent-Infant Interactions: A Systematic Review," *International Journal of Nursing Science* 6 (2019): 445–53.

15. H. Fisher, "Casual Sex May Be Improving America's Marriages," Nautilus, March 2, 2015, https://nautil.us/casual-sex-is-improving-americas -marriages-3002/.

16. M. S. Carmichael et al., "Relationships Among Cardiovascular, Muscular, and Oxytocin Responses During Human Sexual Activity," *Archives of Sexual Behavior* 23 (1994): 59–79.

17. B. Behnia et al., "Differential Effects of Intranasal Oxytocin on Sexual Experiences and Partner Interactions in Couples," *Hormones and Behavior* 65 (2014): 308–18.

18. M. R. Murphy et al., "Naloxone Inhibits Oxytocin Release at Orgasm in Man," *Journal of Clinical Endocrinology and Metabolism* 71 (1990): 1056–58.

19. I. Schneiderman et al., "Oxytocin During the Initial Stages of Romantic Attachment: Relations to Couples' Interactive Reciprocity," *Psychoneuroendocrinology* 37 (2012): 1277–85.

20. D. S. Quintana et al., "Advances in the Field of Intranasal Oxytocin Research: Lessons Learned and Future Directions for Clinical Research," *Molecular Psychiatry* 26 (2021): 80–91.

21. C. H. Declerck et al., "A Registered Replication Study on Oxytocin and Trust," *Nature Human Behavior* 4 (2020): 646–55.

22. D. Scheele et al., "Oxytocin Enhances Brain Reward System Responses in Men Viewing the Face of Their Female Partner," *PNAS* 110 (2013): 20308–13.

23. W. Romero-Fernandez et al., "Evidence for the Existence of Dopamine D2-Oxytocin Receptor Heteromers in the Ventral and Dorsal Striatum with Facilitatory Receptor-Receptor Interactions," *Molecular Psychiatry* 18 (2013): 849–50.

24. H. Walum et al., "Genetic Variation in the Vasopressin Receptor 1a Gene (AVPR1A) Associates with Pair-Bonding Behavior in Humans," *PNAS* 105 (2008): 14153–56.

25. B. P. Acevedo et al., "The Neural and Genetic Correlates of Satisfying Sexual Activity in Heterosexual Pair-Bonds," *Brain and Behavior* 9 (2019): e01289, https://doi.org/10.1002/brb3.1289.

26. G. Sadikaj et al., "CD38 Is Associated with Communal Behavior, Partner Perceptions, Affect and Relationship Adjustment in Romantic Relationships," *Scientific Reports* 10 (2020): 12926.

27. A. Bartels and S. Zeki, "The Neural Basis of Romantic Love," *Neuroreport* 11 (2000): 3829–34.

28. B. P. Acevedo et al., "After the Honeymoon: Neural and Genetic Correlates of Romantic Love in Newlywed Marriages," *Frontiers in Psychology* 11 (2020): 634.

29. H. E. Fisher et al., "Reward, Addiction, and Emotion Regulation Systems Associated with Rejection in Love," *Journal of Neurophysiology* 104 (2010): 51–60.

30. A. Najib et al., "Regional Brain Activity in Women Grieving a Romantic Relationship Breakup," *American Journal of Psychiatry* 161 (2004): 2245–56.

31. R. W. Quiroga et al., "Invariant Visual Representation by Single Neurons in the Human Brain," *Nature* 435 (2005): 1102–7.

32. C. S. Carter and S. W. Porges, "The Biochemistry of Love: An Oxytocin Hypothesis," *EMBO Reports* 14 (2013): 12–16.

GLOSSARY

2D:4D ratio: The length of the index finger divided by the length of the ring finger.

AB cells: Neurons within the bed nucleus of the stria terminalis of mice that are selectively activated by the smell of female mice or their urine.

allele: One of two or more possible DNA sequences found at a single location in the genome.

amygdala: A structure in the temporal lobe of the brain that is involved in the processing of emotional states, including sexual arousal.

androgen: Any of the set of hormones, including testosterone, that activate androgen receptors in the brain or body.

androstadienone (AND): A steroid related to testosterone that is present in men's armpit secretions. It may act as a pheromone.

antiandrogens: Compounds that reduce the levels of androgens or that reduce the brain's sensitivity to them.

autogynephilia: In a person born biologically male, sexual arousal by imagining oneself as a woman or progressively taking on female attributes and identity.

axon: A neuron's single output fiber that conveys signals to other neurons or muscles.

base: A single "letter," or nucleotide, in a DNA chain: adenine (A), cytosine (C), guanine (G), or thymine (T). In the DNA double helix, A pairs with C and G pairs with T.

bdelloid rotifer: A class of microscopic invertebrates that reproduce without sex.

bed nucleus of the stria terminalis (BNST): A subcortical structure that, among many other functions, relays olfactory information into the hypothalamus.

body mass index (BMI): A person's weight in kilograms divided by the square of their height in meters, a measure of thinness or obesity.

bourgeonal: An organic compound with a flowery scent. It, or a related compound, is secreted by the ovum and attracts sperm.

bremelanotide: An injectable drug, marketed as Vyleesi, that is FDA-approved for the treatment of hypoactive sexual desire disorder in women.

cabergoline: A drug that lowers the secretion of prolactin.

CD38: A gene that regulates the release of oxytocin.

chemosignals: Substances, including pheromones, that mediate chemical communication between organisms.

chronophilia: Sexual attraction to persons of a specific age range.

commitment: In the triangular theory of love, the cognitive element—the conscious decision to stay together.

companionate love: The aspects of love that maintain a relationship over the long term—the combination of intimacy and commitment.

conditioning: The acquisition or strengthening of a response pattern by repeated association with a reinforcing stimulus.

congenital adrenal hyperplasia (CAH): A genetic condition in which the adrenal glands secrete excessive amounts of male-typical sex steroids (androgens) during prenatal life.

demand–withdrawal: A toxic style of marital communication in which one partner makes demands and the other clams up or ends the conversation.

dopamine: A neuromodulator released by the terminals of several brain pathways, including the pathway running from the ventral tegmental area to the nucleus accumbens.

emotional jealousy: Jealousy triggered by the belief or fact that one's partner has become romantically entangled with a third party.

epigenetic effect: The addition of chemical tags to DNA or histones in such as way as to alter gene expression.

erotic target location error: The incorporation of the target of one's sexual attractions into oneself, as, for example, in heterosexual transvestism.

estratetraenol (EST): A steroid synthesized in the ovaries that is present in women's bodily secretions and urine. It may act as a pheromone.

fetish: Sexual arousal focused on a specific body part, item of clothing, or other object.

flibanserin: A drug, marketed as Addyi, that is FDA-approved for the treatment of hypoactive sexual desire disorder in women.

frontal lobe: The front-most lobe of the cerebral hemisphere; it controls bodily movements and (in the prefrontal cortex) makes executive decisions about the performance or inhibition of behaviors.

gene expression: The activity of a gene: its transcription into RNA and downstream effects on protein synthesis and structural and functional development.

heritability: The fraction or percentage of the variability in a trait in a certain population that can be attributed to genetic variability within that population.

heterosexual transvestism: Sexual arousal in a heterosexual person from wearing clothes typically worn by the other sex.

histones: Proteins that bind to DNA, maintain chromosomal structure, and influence gene expression.

homogamy: The tendency for people to establish live-in relationships or marriages with partners similar to themselves.

hypoactive sexual desire disorder (HSDD): A lack of sexual desire when the lack is experienced as distressing.

hypothalamus: A small region at the brain's undersurface that regulates numerous goal-seeking behaviors, including male- and female-typical sexual behaviors. It also controls the secretion of hormones by the pituitary gland.

INAH-3: A cell group in the media preoptic area of humans that is, on average, larger in males than in females. It is probably the equivalent of sexually dimorphic nucleus of the preoptic area (SDN-POA) in other mammals.

inferior frontal gyrus: A region within the prefrontal cortex that regulates sexual behaviors in their social context.

intimacy: In the triangular theory of love, a strong liking based on self-disclosure and deep familiarity.

lifespan integration: A type of therapy for rape trauma syndrome that attempts to distance the survivor from the rape by expanding their sense of the intervening time span.

lordosis: In female rodents, an inverse arching of the back that exposes the vulva and permits mating.

major histocompatibility complex (MHC): A set of genes involved in the immunological distinction between self and nonself.

medial preoptic area (MPA): A region at the front of the hypothalamus that, in males, controls male-typical sexual behaviors. In females it may be involved in maternal behaviors. It includes a sexually dimorphic cell group known in humans as INAH3.

meta-analysis: A formal statistical procedure for combining the results of multiple quantitative studies.

microsatellite DNA: A short sequence of DNA bases that is repeated several times; the number of repeats often varies between individuals.

minor-attracted person: Someone who is sexually attracted to minors—a broader category than pedophile.

monoamine oxidase type A (MAOA): An enzyme that breaks down serotonin, dopamine, and norepinephrine in the brain.

moral incongruence: Conflict between one's desires or behaviors and one's moral beliefs.

mutation: A change to the DNA sequence of a genome.

neuromodulator: A neuron-to-neuron signaling molecule that is released from nerve terminals and diffuses within an extended region rather than being confined to an individual synaptic contact like a classical neurotransmitter.

nucleus accumbens: A brain region in front of the hypothalamus that is concerned with reward and pair bonding.

nucleus: In neuroanatomy, a consistently recognizable cluster of neurons with characteristic structure, connections, and functions.

occipital lobe: The rearmost lobe of the cerebral hemisphere, it is concerned largely with vision.

oxytocin: A small peptide synthesized in the hypothalamus. It acts as a hormone and also as a neuromodulator in the brain, playing a role in pair bonding.

paraphilia: An unusual and persistent sexual desire or behavior.

paraphilic disorder: A paraphilia that causes distress or harms others.

paraventricular nucleus: One of the groups of neurons in the hypothalamus that synthesize oxytocin and vasopressin and transport them to their terminals in other regions of the brain and the pituitary gland.

parthenogenesis: Production of offspring by a female without a genetic contribution from a male.

passionate love: In the triangular theory of love, the yearning for union with the beloved; being head over heels in love.

pedophilia: Sexual attraction to prepubescent children or children in early puberty.

pelvic organ-stimulating center: A neuronal group on the left side of the pons that is involved in the triggering of orgasm. The same cell group on the right side triggers urination.

peptide: A chain of amino acids, similar to a protein but shorter.

pheromones: Chemical signals released by one animal that affect the sexual or other behavior of other animals of the same species.

pons: The expanded section of the brain stem between the midbrain and the medulla.

posterior cingulate cortex: A region of the cerebral cortex that, among other functions, participates in the processing of negative emotions such as pain and anxiety.

posterior division of the amygdala: A portion of the amygdala that helps generate sex-specific mating or aggressive behaviors by male mice toward other mice.

prefrontal cortex: The front-most region of the frontal lobe; it mediates higher-order executive functions.

prolactin: A hormone secreted by the pituitary gland that promotes breast development and milk production. It also appears to dampen sexual responsiveness after orgasm.

prostate-specific antigen (PSA): An enzyme present in the secretions of the prostate gland in men and the paraurethral glands in women. It breaks down semenogelin after ejaculation.

rape trauma syndrome: A form of posttraumatic stress disorder triggered by rape or sexual assault.

rs3796863: A single nucleotide polymorphism within the CD38 gene. Different alleles of this SNP are associated with the differences in the intimacy of sexual behavior.

rs53576: A single nucleotide polymorphism in the gene for the human oxytocin receptor. Different alleles and allelic combinations of this SNP are associated with sexual satisfaction within a pair bond.

semenogelin: A protein secreted by the seminal vesicles that immobilizes sperm. It is broken down by prostate-specific antigen.

seminal emission: The loading of the components of semen into the urethra immediately prior to ejaculation.

seminal vesicles: Glands located near the prostate that contribute most of the volume of the semen, including the protein semenogelin.

sexual jealousy: Jealousy triggered by the belief or fact that one's partner is having sex with a third party.

sexual orientation: The spectrum of sexual attractions to males, females, or both sexes.

sexually dimorphic nucleus of the preoptic area (SDN-POA) A cell group in the medial preoptic area of the hypothalamus of mammals that is larger in males than in females.

sexually dimorphic: Differing in size or appearance between females and males.

single nucleotide polymorphism (SNP): A location in the genome where a single nucleotide or base can vary between individuals; the differences are known as alleles.

temporal lobe: The side lobe of the cerebral hemisphere: It includes cortical areas concerned with vision and memory, as well as subcortical structures such as the amygdala.

transgenic: Describing an organism, such as a mouse, to which a foreign gene has been added.

triangular theory of love: Robert Sternberg's theory of love, according to which love comprises three elements—passion, intimacy, and commitment—combined in variable proportions.

vas deferens: The pair of tubes that transport sperm to the urethra.

vasopressin: A small peptide related to oxytocin. It acts as a hormone regulating blood pressure and as a neuromodulator in the brain, contributing to pair bonding in some species.

ventral pallidum: A brain region adjoining the nucleus accumbens that, like the accumbens, is involved in the processing of reward.

ventral tegmental area (VTA): A region within the midbrain. It includes neurons whose axons pass to the nucleus accumbens and prefrontal cortex and release dopamine there.

ventromedial nucleus (VMN): A nucleus within the hypothalamus that regulates female-typical sexual behaviors as well as aggressive behaviors by both sexes.

vomeronasal organ (VNO): A sensory structure within the nasal cavity of some species that detects chemosignals.

Westermarck effect: The phenomenon whereby children who are reared together rarely experience sexual attraction to each other in adult life.

INDEX

McPhail, Beverly, 200
McPhail, Ian, 173
medial preoptic area (MPA), 50–51; in
    mouse, 57
Mellor, Emily, 186
memory, 48
menstrual cycle: sexual attraction
    around, 29–31; sexual desire
    around, 31
menstrual synchrony, 31
Meselson, Matt, 17
Meston, Cindy, 1–2, 4–5, 19, 21, 118, 123
mice, hypothalamus in, 55–60
minor-attracted person (MAP), 173,
    175–176
Monti-Bloch, Louis, 36
moral incongruence, 190–191
Morrison, Evan, 24
mother-child attachment, 230
mothers I'd like to fuck (MILF), 38
mud snail, New Zealand, 5–8
Müller, Karolina, 171–172

Najib, Arif, 238
Nelson, Lief, 24
neurohormonal theory of sexual
    orientation, 84–85, 88, 90–92, 95
neurogenetics, 56–60, 62
Nickerson, Todd, 173–176
NoFap, 190–191
nucleus accumbens, 51, 52, 238

Oakley, Susan, 100
occipital lobe, 43. See also visual
    cortex
OKCupid, 38
Olfaction: role in arousal, 54–55; role in
    sexual attraction, 33–36
olfactory bulb, 55
Olver, Mark, 173
one-night stand as precursor to long-
    term relationship, 231

online dating sites, 127–129
orangutans, 208
orgasm: cervical, 103–104, 111; clitoral,
    99–102; G-spot and, 102–103;
    multiple, 108–109; prolactin and,
    108–109; prostatic, 110–111; role of
    pons in, 61; and transsexuality, 107;
    vaginal, 102; in women and men,
    106–107
orientation, sexual. See sexual
    orientation
oxytocin receptors: in humans, 233, 234,
    in voles, 223, 225–226
oxytocin, 60; during orgasm, 231; effects
    of administration, 232–233; in new
    lovers, 231–232; role in love, 230–235;
    role in pair bonding, 221–226

Pace, Peggy, 215
pair bonding: in birds, 220; in prairie
    voles, 219–226; role of oxytocin
    and vasopressin in, 221–226; in titi
    monkeys, 229–230
palindromic DNA, 15
Palmer, Craig, 207–209
paraphilias, 149–162; and brain injury,
    155; causes of, 152–155; conditioning
    in, 152–154; definition of, 149–150;
    genetic factors in, 154; lack of social
    skills, 154; number of, 150; paraphilic
    disorder, 150; in women, 151–152.
    See also acrotomophilia;
    autogynephilia; BDSM;
    chrematistophilia; fetishism; furries;
    internalization of erotic targets;
    pedophilia; sex with animals; rubber
    fetish; stocking fetish; voyeurism
paraurethral glands, 102, 106, 111, 114
parthenogenesis. See asexual
    reproduction
passionate love, 227–228
Paul, Thomas, 87

Printed in the USA
CPSIA information can be obtained
at www.ICGtesting.com
LVHW091500021123
762748LV00020B/351/J